SOUND

ALSO BY BELLA BATHURST

The Bicycle Book
The Lighthouse Stevensons
Special
The Wreckers

SOUND

A MEMOIR
OF HEARING
LOST
AND FOUND

BELLA BATHURST

GREYSTONE BOOKS
Vancouver/Berkeley

Greystone Books Ltd.
greystonebooks.com

Cataloguing data available from Library and Archives Canada
ISBN 978-1-77164-382-5 (pbk.)
ISBN 978-1-77164-383-2 (epub)

Cover and text design by Peter Dyer
Cover photograph by Mike Hewitt/Getty Images
Printed and bound in Canada on ancient-forest-friendly paper by Friesens

Greystone Books gratefully acknowledges the Musqueam, Squamish, and Tsleil-Waututh peoples on whose land our office is located.

Greystone Books thanks the Canada Council for the Arts, the British Columbia Arts Council, the Province of British Columbia through the Book Publishing Tax Credit, and the Government of Canada for supporting our publishing activities.

Canadä

For Urièle and Marc, with love

In the 1981 census, a return was received from someone who described their profession as 'sculptor of stone lions'. In answer to the supplementary question 'Describe what your job entails,' the responder wrote: 'I chip away all the stone which is not lion.'

Contents

1

Sailing

I TAKE THE TURNING off the main route, slowing to follow the single-track line of tarmac out towards the sea. It's late summer and the verges are thick with bracken and fireweed, their stems knocking at the car as it passes. Over to the north the sky has blackened, banks of cloud massing. Half a week of perfect weather, and it looks as if it's going to break now.

Six miles on the track widens. There's a farm, a scatter of buildings, an office. Beside the verge lie the curvaceous outlines of several upturned canoes with a rack of yellow kayaks poking out of the car park beside it. As I pass the farm a collie comes hurtling out of the yard, racing towards the car. He flings himself at the bonnet and in the instant of his collision I flinch the steering wheel away. His fur flattens against the glass and I feel the thud of his body echo through the metal. A blur of hackles and then he's gone, slid out of view. *Oh God, have I run him over?* Then he's up again, a snarl of black and white against the closed window. His tail is up and his teeth are bare and there's so much energy in him, so much unleashed delight in his rage. He jolts up, front claws ticking on the glass. I can see the force of each bark shoving his whole body forward: thud, thud, thud. The slower I go, the angrier he gets.

I straighten the steering wheel and drive on, the dog receding in the mirror. He's running down the track, the tuft of his tail peacocking his achievement. After a few moments he stops and watches the back of me.

Just beyond the farmyard there's a passing place so I pull in and stop the car. I look down at my hands, watching them shake. It's not the dog. It gave me a fright, but dogs love chasing cars and that collie has done it before. It was the fact that I couldn't hear the dog. Nothing. Not a whisper. He was giving it his all, every last atom in him, and none of it reached me. No sound, just a sort of muffled rush, and even that might have been my imagination.

I put my hand up to my right ear and cover the little hole on the right hearing aid. Open, close. Close, open. All working fine and the other one too. The aids are small plastic flesh-grey in-the-ear things designed to fit down the tunnels to my eardrums. At the outer end there's a socket for the battery and a tiny gap for sound to get in. At the inner there's another slot where the processed sound enters my ear. I take them both out and dip my head from side to side, then turn the radio on and off. There's nothing blocked, nothing out of the ordinary.

I put the aids back in and the world returns. These ones are designed to cope best in indoor environments, places like offices and homes full of double-glazing and other human beings all soaking up sound. The aids don't deal well with big reverberant spaces and they really don't seem to have got the hang of the outdoors at all. Out here the wind works on the aids the same way it does on a phone, drowning voices below a storm of white noise. If I turn my head slightly the noise eases but that probably means turning away from the speaker, and I need to see people properly in order to be able to hear them because half of what I'm listening to is the words on their faces. Take away most of the sentence, and I can still see whether it's a question or a statement, or pick up enough of the tone to gauge the mood of the speaker. I'm looking for physical guides, and half their diction is there in the features – the lift of an eyebrow, a hardening tone, the warmth in the eyes.

At this time, in 2004, I am deaf. Not completely deaf, just down to about 30 per cent of normal hearing. I had started to lose my hearing in both ears about seven years ago and it has been declining ever since. I wear hearing aids in both ears and when I take them out, I can hear individual fragments of sound but not really the links between them. Certain words in a sentence or specific sounds are audible, but music is only a beat and a voice is just a chain of broken plosives.

I am here because I'm going sailing. My friend Eric has asked me to come and I very much want to do this. But water is not my element. I'm mesmerised by it and I'm scared of it, and because I'm scared of it, I head straight for it. I love sailing – the adventure, the pleasure of being with friends. But I hate sailing – the cold, the wet, the seasickness, the fear. I'm here because I want to push the love in and the fear away, and at that time I believe the best way to deal with fear is to hurl myself head first at the thing which most frightens me. It is a kind of idiot courage, a determination to force myself towards a different shape. I hope if I do the things that scare me for long enough, they will become easier. I want to override the physical facts, and to teach myself a lesson.

The trouble is that water is not like land. It's said that almost all of what we experience as sound is an echo – that the conversation you hear when you're sitting in a room with someone is mostly just their voice bouncing off the walls. Outside, at sea, there's very little echo. Still water carries sound beautifully because there's such a big surface area from which it can rebound, but choppy water presents a thousand different points of connection. The only surfaces with a good echo are the ones on the boat itself – fibreglass, metal, wood. And when the engine is running it lays a steady thrum over any lower sounds. What that means in practice is that all the sibilants get knocked out of speech: Nyoupuen-erou? Ucuthemooinroe? Ilanacoupleougarleae?

The difficulty is that sailing requires a lot of instructions and understanding. In the heat of the moment I can't expect anyone to stop what they're doing and turn towards me so I know the next couple of days are going to be tricky. If I can't hear something on land it's a problem but there are things I can do to improve it. Out here, blanking an instruction or mishearing a command has much bigger consequences.

I start the car and drive on. Half a mile beyond the farm, the road opens out to a view of water. Down the slip to the loch there's a trailer, a tangle of broken ropes and a few fading fishing nets. Three lines of boats lie anchored in the bay laid out with their bows to the land and their sterns pointing towards the weather. I park up and sit for a second, eyes closed. The bracken is high and there's a bramble patch directly in front of the windscreen. The berries are ripening and when I look at them I feel a surge of affection for all this abundance, the trees and the moss and the solid ripe comfort of earth. I'm homesick for land, I realise, and I haven't yet set foot on the boat. Beyond the windscreen the trees have stilled. No movement from the birds. The loch is clear and a bee probing one of the late foxgloves vanishes into the top blossom. Just for a second time stands and waits. No sound, no movement, just a single moment suspended.

I inhale a single lungful of breath, climb out, lock up, and head down the track. At the top of the pontoon is a car and two people unloading bags. One small boyish figure staggering out of sight down the walkway, and one larger figure, hauling a bulk of sailcloth from the back.

The big figure sees me. He puts down the sail and as he straightens up his smile is wide. All the pleasure of a proper adventure in there, three whole days filled with nothing but the pursuit of fun.

'No Tom yet,' says Eric, hugging me.

I pull my jacket up above my head, grab the bags and scurry down the walkway towards *Lismore*. There she is, surrounded by bags and ropes and lines, primed for a great escape.

Luke, Eric's nine-year-old son, climbs up the hatchway. 'Hello, Bella!' he says as Eric and I lift bags onto the foredeck. 'Dad? There's a drip.'

'Where?'

'In my room,' says Luke.

'Forepeak,' says Eric.

'My sleeping bag's wet.'

Eric swings himself over the side and goes below. In the main cabin there's an overwhelming scent of diesel, old milk, silicone sealant and mould. For the next half hour or so, the three of us gaffer-tape leaks and unload supplies, fill the lockers with food and bounce our sleeping bags down through the hatches. Eric checks the fuel and fresh water lines and looks at the safety gear.

It's been a patchy summer. Most of the time *Lismore* just sits here, looping round her mooring buoy to the motion of the tide and the current. But West Coast weather can be capricious and back in May she got badly beaten up in a gale during which several other boats broke loose from their moorings and set off on an all-night wrecking spree round the loch. Though her chain held, *Lismore* was hit by a couple of rogue boats and ended up with a damaged steering mechanism. This will be the first longish trip she's been on since then, which is why Eric has chosen a gentle inshore circumnavigation of Mull.

By the time we're ready the rain has passed over. The tide is slack and the water is dark, its surface serrated by ripples. We should get going. Just as Eric empties a packet of crisps into a bowl a car appears near the top of the pontoon and a man comes stumping down the walkway. He's wearing shorts, an overcoat and a pair of bright orange trainers, and he's adjusting the strap of his bag as he talks into his phone.

Tom is Eric's cousin. He and his wife run a hydroelectric company based in Perthshire installing small-scale pipelines all over the Highlands. Before that, he was in the Royal Navy. In fact, Tom is an ex-submariner trained by Her Majesty's Government into a state of extreme competence in any situation from minor catering issue to full-scale nuclear attack. He smiles at us, shoulders a small holdall and a laptop bag onto the deck, and carries on talking. Luke walks over to the bags and looks at them with interest as a heron takes off. It is a further five minutes before Tom ends the call and shoves the phone into his back pocket.

'Hello, lovely people!' he says, somehow managing to kiss each of us in turn whilst also untying the mooring ropes and leaping on board. Eric pushes at the throttle and the black line between land and boat looms wide beneath us.

And we're moving. The tide itself pulls us out on the ebb, though it takes us a further half an hour to make our way from the bay to the sea. Eric stands at the wheel while Tom rides the buoys out into deeper water. By the time we round the final bend and see the Firth of Lorn wide in front of us it's nearly 5 p.m. As we reach open water, the breeze gets breezier and the water gets choppier. There's no single moment when land becomes sea, just a gradual decrease of one element and increase of another.

'Right,' says Eric, looking at Tom and then at me. 'Ready?'

We need to raise the main sail, the hump of cloth currently folded down and tucked along the boom. Tom and I walk along the deck and brace ourselves against the mast. I can hear Tom saying something, but all I can see of him is a pair of feet. I look over at Eric. He nods. Attached to the mast are a selection of different-coloured ropes, none of which have any obvious purpose. Tom says something. I can't hear, so I twist round the mast.

He points. 'Uncleat the halyard.'

I look at him, blank.

'This.' He taps a green rope, unwinds it and begins to pull. I return to my side of the mast, unwind its equivalent, and begin hauling. Tom's legs strain against the weight as the great white shape snags its way upwards. Once it's high enough, we can use the winch levers to raise the rest. I follow Tom's lead, watching his movements until the sail is tight against the top of the mast.

Eric shuts off the motor. In the silence left by the engine the boat's motion changes. Instead of moving forward smooth and upright, it's now directed by the wind. The deck starts to tip to the right and the looped ropes curve out while a coffee mug on the back locker rolls into the cockpit. When the motor's running I can feel its vibration through my feet and hear its varying pitches low in my ear. Without it, wider sensations rush in – a slur of voices, the trace of a tail-note from an oystercatcher overhead.

And then somehow we're sailing, the boat pulled along by nothing but air. I look up at the mast, now twenty degrees off the vertical. Tied high up on the mast are Luke's bicycle and two flags, the saltire and a pirates' Jolly Roger. Both are streaming out behind us now, stiff with the force of the wind. The shush of water over the hull makes it seem as if we're racing along. The sails are taut and the telltales are rigid, the wind's in our hair and the coffee has hit the right bits of our systems. The scatter of nearby yachts somehow makes the water seem benign, even joyous.

On land the light lengthens. There's rain over the hills and a long flat bank of cloud over to the south while the wind from the south pushes us up towards the Sound of Mull. For a few minutes we can relax. We sit on the benches in the cockpit holding tea or beer, watching the water do its work. We chat, filling each other in on work, kids, relationships, filling the gaps left by several months' absence. Eric is just about to head off to Sudan for a month on a work trip, Tom is having trouble with a non-paying client, I've just moved house.

This is fine, I think. It's going to be fine. I can hear them. Sitting here, things are easier because I can see their faces clearly and it's daylight and the wind can't get at us so easily when we're sat down in the cockpit. Through the fibre of the hull I can feel the sounds of the boat itself – the slap of the halyards, the deeper, more descriptive sound of the wind against the boom, the sailcloth tightening or loosening depending on our speed and direction. And under it all the susurration of the sea itself. The shush it makes as it slides along the hull, fast or slow, urgent or gentle, its mesmerising endlessness.

One of the benefits of deafness is that it teaches you a lot about acoustics. Pilots talk about air as a visible thing; its speed, its flow, the quirks and currents in a particular valley, the way the wind assumes a colour or mood. The same applies to sound and the way it's shaped. Cold means more reverberance, hot means less. Although it's warmer, city air is harder to hear than the air in the country. Snow softens sound though the cold in ice clarifies it. Out here, the air is cool, but the boat itself contains a series of different acoustic microclimates – the cockpit, the deck, the galley.

And there's also the universal human ability to tune in and out of other people. Everyone has it and everyone does it, every day, all the time. Even if you haven't seen an old friend for a long time their voice stays within you – the way they speak, their rhythm and diction. Far back down the side streets of the human brain is a tiny recording studio in which all the tracks of a lifetime are laid down: the difference between a true and a false laugh, the exact way your husband says 'brilliant' or 'orchestra', your colleague's habit of elongating 'Glasgow' but not 'glass'. Without acknowledgement or effort, you can recall exactly the way your son's first cries differed from your daughter's or the way your brother drops an octave when he's trying to impress someone. You know what your boss sounds like when she's nervous and how that differs from the way she speaks with friends. You could calibrate the note of danger in your partner's voice down to micro-decimal points or discriminate between several thousand different shades of 'no'. You know what your girlfriend sounds like when she's flirting, or how when she's angry the Geordie in her comes to the front. You have the world's most perfectly engineered voice-recognition software invisible, inside.

In fact, it doesn't even have to be people you know. If you turn on the radio you know who's speaking without the need for introductions. You understand which politicians' voices sound honest and whose pitch you'd never trust. Without knowing it, you've already spent half a lifetime familiarising yourself with total strangers. You can hear all the altered shades of mood between one day and another or pick out the difference between rehearsed and true. You can locate the pinch of fear, the ease of lies or the warmth of laughter. It's all there, all available. All yours.

And I know these three voices well. Eric has been skippering a boat for so long he could probably be heard in Rockall. Nice clear voice, doesn't slur or run his words together. When he first bought *Lismore* he made a vow never to become a shouty skipper because he'd already seen enough people terrified by some old Ahab who transformed from easygoing charmer to maritime psycho once on deck, so he never yells. He's learned well enough to throw his voice from bow to stern and get stuff done without the need for keel-hauling. Confident diction, opens his mouth properly as he's speaking, doesn't nibble his sentence ends. Luke speaks just as clearly as his dad though his voice is higher and thinner and what he wants to say sometimes comes out in a rush, so I have to concentrate much harder on reading his face.

Tom is a slightly different case. Because of his military

training, Tom does everything in life while maintaining exactly the same tone throughout: agreeable, instructive, aimed always at a level of understanding somewhere between cabinet minister and small child. It is a steady, patient voice designed to convey both authority and a lifetime spent dealing with people just marginally less magnificent than himself, though his natural pitch is very low. On land he could hold the Albert Hall, but out at sea his words have a tendency to sink beneath the waves. He's used to dealing with that by slowing down, but, though his diction is perfect, I can't get away with not seeing his face.

But so far, I tell myself, I'm on the boat and we're outside and I can still hear them all. Everything will be OK. It will all be absolutely fine. I don't want to tell them that my hearing seems to be getting worse because saying it out loud just makes it twice as true. And I'm sure it's unnecessary – these are people I love and want to live up to, so I don't want to make some victimy grandstanding disability-awareness statement.

It does not occur to me until a very long time later that, by failing to explain how bad it has got, I am making life more difficult for everyone else. And that, in trying to take up less space on the boat, I am actually taking up more.

By 6 p.m., the rain has passed over us. Tom is steering and Eric is over by the stern helping Luke reel in the fishing line. I am down in the galley making tea. Tom is making calls and Eric has got his back to me. I'm concentrating on the stove, and for a couple of minutes I hear nothing. Eric appears in the hatchway, blocking the light. He's pointing urgently towards the back of the chart table.

'... forecast! Radio!'

The evening shipping forecast, essential to all sailors. It's a VHF set, not straightforward, and I'm fumbling, unable to hear more than a crumple of static. I don't know how it works so I keep turning back to face Eric because he's behind me. It's taking too long. He lifts himself over both me and the cooking in one practised motion, lands by the chart table and adjusts the set.

At home, the shipping forecast sounds like a landlubber's prayer of thanks for not being at sea. At sea, it sounds like a different kind of prayer entirely. Barometric pressure, visibility, Beaufort Scale, the rise and fall in rhythm like the swell of the tide. But all that poetry is just rococco over the real story: bad weather with some stupendously bad weather to follow. As I stand close to the set I can just about understand that there's an area of low pressure in the Atlantic heading in from the west, strong winds from the south and maybe half a day of mixed fortunes. Tomorrow morning might be OK but the rest sounds like trouble.

'I can see rocks!' shouts Luke from the bow. He sounds delighted.

It takes us half an hour to get ourselves through the entrance to Loch Aline. By the time we start looking for a mooring buoy, the light is almost gone. Along one side, there's a slab of forestry. On the other there is dust, several acres of raw quarried rock, a jetty with a couple of unladen cargo ships moored alongside. No buoys.

We need to get the anchor out so Tom and I head to the front and begin trying to disentangle the chain. It's snagged halfway along its length. One of the links must have picked up a stone and fouled itself on the links farther down. Both of us are pulling but the stone won't come free.

'... n ... ou ... allet?' asks Tom, head down, his sentence vanishing below the chink.

Sorry?

'Hammer,' he says. '... thi ... o ... t it wi ...'

Hammer. I walk back over to the locker by Eric and start going through it, looking for the right tool.

'Hit it!' says Eric, his voice raised. 'Hit it on the capstan!' He mimes.

What's a capstan? Back along the deck, where Tom is

already banging the snagged link against the winch, pushing the wedged stone out. As the chain finally comes free Eric puts the boat into neutral and then reverse. As we release the links down into the water I feel rather than hear its weight through the soles of my feet. The water slides out backwards along the bow. Nothing. The anchor still hasn't caught. Again we pull it up and again we lower it. It takes us three more tries before we're fixed, held tight in seven metres of water. By now it's dark and there's a mean little breeze stealing the warmth from the backs of our necks. Luke is below, eating pasta and cheese.

By the time we've got things together and Eric has made supper, it's too cold to sit outside. Excellent, I think; that means at least 20 per cent fewer 'sorry?'s. There's light down in the diesely fug of the saloon. Outside in the night there's only the electronic glow of mobiles to illuminate their faces, which means I'm more likely to mishear. I don't like having to ask people to repeat. I find it annoying and I assume others must too. *Swing over to walk on the right side, look round from the washing up, read the reflection in the window, strike for the one clear word in the sentence.*

It takes us most of the following morning to motor up the Sound of Mull, sitting in the sun, eyes closed, not enough wind even to attempt sailing. The Sound is quiet for this time of year, mostly just lobster boats and ferries, plus a few yachts sliding in and out of Tobermory.

Eric hands me up a cup of coffee, climbs out of the hatch and starts examining the electronics around the wheel. 'Bloody autopilot. Give me some throttle?'

At the base of the wheel is a gear lever, nice and self-explanatory: forward for forward, backward for reverse, centre for neutral. I push the lever forward and Eric takes his hands off the wheel. *Lismore* pulls to the left, sending us round in a broad circle. Luke and I stand by the prow and watch the far end of Ardnamurchan rotate past us three times. 'Da-ad! What have you done?!'

'Arse!' says Eric. 'Bastard! Bastard! I fixed that thing! Sorry, Luke. I fixed it, and it worked fine.' He disengages the mechanism, takes the casing off, examines it, replaces the casing, steps back and kicks the wheel base twice.

Same thing. The faster we go the tighter the loop. After twenty minutes, we're all dizzy. Eric picks up the chart and removes a fragment of toast from the top of Ben More. 'OK, we'll just have to have someone on the helm the whole time.'

We spend the morning on the island of Lunga, chatting, eating, watching puffins, forgetting the time. We haven't been back on the boat for long before the weather starts to shift. From the north-west, a flat covering of cloud comes sliding in over us. It's grey and featureless and it's travelling fast, blocking the light and rendering everything down to a shadowless flat. There's a sense of weight and purpose behind it, and of the sea's temper rising. We need to get through the Sound of Iona and round the south-western point of Mull to the mooring at Ardalanish tonight. The early part we can probably sail but the rest we'll have to motor.

By the time we're parallel with the abbey on Iona, the feel of the afternoon has changed completely. The wind is up and the current is running strong towards the main-land. The temperature has dropped and all of us have put on extra layers of clothing. The water in the Sound is restive with an odd uneven swell to it. The last few tourists on Iona are making their way through the gravestones down to the shore, waiting for the ferry.

As we pass the workmen's cottages on Earraid I'm kneel-ing on the back locker making calls when the wake from a passing RIB hits us. A gout of water hauls up over the transom, hitting the guard rail and soaking me. When I wipe the water off I realise that the mobile is OK but my right hearing aid has gone dead. Oh God. I take it out and try to revive it. Nothing. *Damn*. Damn for several reasons. Damn because the right ear is my good ear. Damn because although these are insured I only have the replacement for the left ear here on the boat with me. It looks like I'll just have to work with one ear.

By the time we emerge through the southern side to the Sound and turn the corner back towards the mainland, the light has shaded over into a sullen dusk and the two or three other yachts near us have fled for Tobermory, sails down. I watch Eric's face for cues. He looks purposeful and focused, concentrating on the currents. Good. That's good, I think.

In order to get round to Ardalanish we have to pass through the gap between the edge of Mull and the Torran Rocks. The gap between the rocks and Mull itself is a mile wide and though there's now a tide doing its best to drag us inland, there's also a wind pulling us out. The main difficulty with the Torrans is not the rocks themselves but the weather system those rocks create. Far beyond their extremities there are rumples and boilings in unexpected places and currents where none should be. This is not the smooth-skinned sea we were looking at earlier this afternoon from Lunga, this is a darker character full of conflict and haunts. Even now on the approach, there are flat patches surrounding us, brazen circles where the water has been hammered out like flattened tin. When we move through these circles, there's a thumping sensation as if something – or someone – had just bumped along the hull beneath us.

The best thing to do is to hug the coast of Mull, creep round the Torrans and hope really hard that the rocks don't see us. But here down the line from the Iona side comes the wind. We see it before we feel it, a darkening on the water, a blue-grey presence opening and spreading wider and wider until it chases all the sea down dark before it. The sheep on the Mull side have gone into one of their foul-weather huddles braced against the rain, and when the wind hits us it pulls at the wires so hard the note thrums all the way through the hull.

It is just as we reach the eastern point of the Torrans opposite the tiny islet of Eilean a'Chalmain that the engine fails.

The first we know of it is a break in the rhythm. Then the boat gives a snort of surprise, the rhythm restarts for a couple of seconds and, with a last leap, it's over.

Luke is sitting in the galley, slice of pizza paused in his hand. 'Dad?'

Eric looks at Luke and at Tom, who is over by the mast on his phone. Then he thrusts the wheel at me.

Tom has already ended the call. As Eric starts undoing the bindings on the sail, he calls something over his shoulder in my direction.

From this position, the wind is blowing directly into the left hearing aid, obliterating everything. I turn my head and yell, 'Can't hear!'

Again Eric shouts and points. Again I can't hear. Holding the wheel with one hand, I climb up onto one of the lockers, trying to see more of his face. It takes three tries before I get what he's saying.

'Two hundred and eighty degrees.' He points over his shoulder south-east towards the Firth of Lorn. 'Over there! Hold it steady till we get this up!'

At the moment, we're pointing towards Mull. I look at the horizon, then at the different electronic readings, then at the water. The wheel is wide, requiring both hands to span it, and somewhere through its inner workings I can feel the rudder cutting through the water. Unlike with a car wheel there's a delay between the turn of the wheel and the response from the boat. I'm not confident at helming and the confidence I do have always vanishes into that little pause. *Why doesn't it respond instantly? What if it's supposed to turn the other way?*

As Eric and Tom concentrate on the sails the boat slides leftwards into one of the circular boilings. Almost immediately, the wheel goes slack. The rudder has lost traction in the water and the wheel just whirls in my hands. I turn it from side to side but there's nothing – no feel to it, no grip on the water.

It takes me two or three tries, hauling the wheel from side to side, before I can get us anywhere near 280 degrees. Almost as soon as I do so it slides out again, jibing to the left. We're circling now, the coast of Mull slipping past us. It doesn't matter which way I spin, the rudder still feels as light and aimless as if it's broken. On either side of the mast, Tom and Eric haul.

Eric yells something. I can't hear him. He yells again. I still can't hear him.

I bend down and call through the hatchway, 'Luke? Can you come up and tell me what Eric is saying?'

Luke looks up. But as he walks up the steps, he's barefoot and wearing a onesy with a fleece on top. I hadn't realised, but he'd changed for bed an hour ago. There's no way he can go on deck wearing that; he can't clip on.

The sail is halfway up and I can see Eric straining, all his weight at the winch. I can't get this boat to move, and nor can I get it to stay still. As the sail rises, the boat begins to heel over. Still nothing. We're still pointing in the wrong direction and the water is still controlling us. There's so much water on my glasses I should take them off and wipe them, but I need both hands. I slide the wheel round again to the Mull side and this time feel the beginnings of resistance. Up above, the sail cracks hard in the wind. I keep turning, heading towards that resistance, understanding just enough to know that the wind will only fill the sail and drive the boat when it's at certain angles. When the boat has turned to a point where the wind spills out the sail goes slack, the boom swings round and all that weight of sailcloth slaps around uselessly. The trouble is that the angle where there's most resistance is not the direction we should be facing.

Just as I sense the beginnings of connection, we slide into another boiling. As it turns and slackens, the sail empties and the boom swings round again. The sails slap, a great panicky battering against the mast: *BANG! BANG! BANG!* Oh God, I can hear that all right. How can it make so much *noise*? Both Eric and Tom have stopped trying to raise the sail and are yelling something at me. That way! *That way!* I stand there, knuckles white, spinning this stupid directionless wheel in the middle of this old wet graveyard tuned to the will of the sea.

Through the rain I can half-see Tom shouting the shape of my name, but the gale takes his words and flings them overboard. I can't turn out of the wind because if I do I can't see Eric or Tom so I just stand there, glaring over the edge of the spray hood for some sort of clue.

Tom is gesticulating, pointing up the mast at the tell-tales, the little bits of string hanging off the top of the mast which act as guides to wind direction, and then at something behind me. Eric has one hand cupped around his mouth and is hauling at the ropes with the other.

Tom lets go of his rope and strides down the deck, pointing behind me. The half-risen sail bags and slaps. 'The Minch! The Minch!'

The *Minch*? The Minch is the stretch of water between the Outer Hebrides and the top of the Scottish mainland. It's miles from here, due north. What does he want the Minch for? He was pointing south-eastwards before. Obediently, I swing the wheel round. That falling pause, then the wind smacks us and the boat tips so sharply there's a slam of falling crockery from below. All three of us on deck have to grab for handholds and, when I look up, I see that we are now beam-on to the current. Beam-on, I think: not good.

Both men stop winching and stare over. With his free hand, Tom is making exaggerated spinning motions. 'Round!' he yells. He lets go of the mast and stabs a finger out to sea. '...! ...!!'

Rain spatters against my glasses. I look upwards, then sideways, trying to catch his words. 'That!' he yells, now pointing behind me. 'Mainsheet! *Mainsheet!*'

Mainsheet? Not the Minch. I had misheard. I look behind me and then up again at Tom. 'The mainsheet! Undo the mainsheet!'

Sheet? *Sheet?* That's not a sheet, it's a rope. I spin the wheel back to where I was trying to get it before and yank the end of the rope out of its cleat. We're turning again. *No grip.* Is it my imagination or do the Torrans look a lot closer? The sheet/rope comes free and the boom swings round to the right with a bang. I pull it taut and look back at Tom. He nods, so I yank it back into the cleat and tie it off again.

Luke is standing on the step at the top of the hatchway, peeping over the top of the hood. 'Luke,' I say in my best totally-no-problem voice, 'maybe best to stay below.'

Luke turns round, screws his face against the rain and vanishes into the cabin. I look at the warm yellow light down below, that little fairytale haven where there's food and heat and the invincible certainty that grown-ups can fix things. Rain drips down my neck.

Again, there's the faint feel of resistance beneath us. I steady the wheel and try to head into that sensation. We're nowhere near 280 degrees, but I'd do anything to get out of this godforsaken water and the overwhelming crime of the sails. Why won't this place let us go? Eric points again, shouts again. I can't hear his voice, but I can see where he's pointing. *Over! Over!* Hard to the right! I turn again, and the resistance disappears.

'... ell ...! ... ll ...! ... t ... ay!' Eric is holding on to the mast with one hand, trying to amplify his words by cupping the other around his mouth. I can't see a thing. I'm too terrified to shrug so I just stand there holding the wheel so tight my fingers cramp white.

He turns back, says something to Tom, jams the winch

back in its hold and strides down the deck. As he gets to me he looks me full in the eye and says clearly, 'Go and help with the sail.'

As he moves to take the wheel, I find my fingers have petrified to the metal. 'Sorry,' I say, unsticking them slowly, not looking.

Up on deck, the sail is up past the halfway point, but the clips linking it to the mast have snagged. Eric angles the boat to the best of the wind, lets it drift towards the edge of one of the boilings and then stops the wheel down hard in one position. This time, instead of turning back to dead water, *Lismore* pushes through and into the current. Beneath me, the sea hisses white. Tom, on the upper side of the deck, is putting his full weight into turning the winch, but the cloth keeps snagging. And every time I look up the Torran Rocks seem to reach a little closer. The wind sings again through the wires, slackening and then shoving hard, and even Eric is having trouble finding the right point of sail.

It takes Tom and me what feels like hours to raise the last few inches, cursing at the winches and stabbing at the clips with the boat hook. I'm leaning backwards at an angle of about 25 degrees and I am shaking, from fear or from cold, I don't know which. When I hear the rumble of Tom's voice I no longer respond. I can't face my own faults in not hearing him again.

It gets dark – a shineless blackening – and it gets cold. All we can see of the Torrans now is an outline against the sky, a thickened shape massing at the edge of our attention every time we turn. Though Eric has got us back into easier water, we can't move past the rocks in the right direction until the sail is fully up and we can start tacking from side to side, stitching a wake over the sea to a place out of danger.

When the sail has risen to within the last few inches of the top, Tom taps my arm and makes a cutting motion. We cleat off the ropes and sag back to the cockpit while the wind circles us, looking for the weakest spot.

It is a further hour and a half before we have tacked our way out of the Torrans' reach. We haul and untangle, unspinning one side to spin the other, waiting till the sail flips behind the mast, too drenched to say anything much.

In between tacks, Eric works at the diesel fuel line while Luke, still in his onesy with a foul-weather jacket and a pair of wellies on top, wobbles a torch above him. There's a lumpy swell rising and the moon is still lost behind clouds.

After an hour, Eric pushes the cover down. Luke clambers onto the chart table and turns the ignition key. There's a short impassive whine and then a cough which reaches right deep down through the whole boat. I hear a choke, then a splutter. The engine shrugs, almost cuts, recovers itself. The third time Luke turns the key, it steadies.

Slowly, gingerly, Eric looks up. He raises a finger, listening. Then, like a parent tiptoeing away from a wakeful baby, he replaces the hatch cover and steps back. He raises both hands, palms up. All of us listen. Through my feet I can feel a catch in the motor's throat, a slip in its rhythm. But it's going. It's going, and we're moving, and that's all that matters.

By the time we've found Ardalanish and dropped the anchor it's almost midnight. The following morning, Tom leaves. He has to get back for meetings and he can catch the ferry over to the mainland from Craignure. The three of us sail back towards the mainland, haul the sail down and motor towards the hidden inlet.

'I'll clear, you helm,' says Eric, handing me the wheel again and holding my eye just a second too long. He wants me to get my nerve back. 'Can you take us in?'

Oh God, I think. I can't do that again. I can't hear, and I can't steer.

'OK,' I say.

To get back to the loch, we need to steer back up the narrow two-mile channel, slaloming our way around the red and green port-and-starboard buoys. The mouth of the channel is wide but then it narrows abruptly, shelving down on one side into a disorder of rocks and spits. Small white cottages and farms line both sides, some obviously locked up for winter. To negotiate the entrance successfully boats have to come right in close to the northern bank, through a strong tidal race, and then swing out again round the corner.

I look at the depth finder and wonder whether it's playing up. One minute it says we've got fifteen feet of water below us, the next we've got two. Some way ahead of us there are two buoys, one red and one green, both almost obscured by white water pushing over their tops. *Red to port, green to starboard.* Except that here, because of the tight corner, green is to port and red is to starboard. I slow down. Bad idea. *Trust what the markers tell you. Rev the motor, pull the boat right in to the very edge of the shoreline, make the corner. Do it fast.* The depth finder is all over the place; ten feet, three feet, twenty feet. The wind is picking up again, the sound of it pushing at the remaining hearing aid.

Eric is sitting on the deck surrounded by folds of sail, the phone in his hand. Down near the water's edge I spot two signs: a black and red stripey pole and a white circle, one about twenty foot from the other. Red and black must be danger but I don't remember anything about white marks. Is it a wreck sign? Twelve feet. Eric looks up, says something. I can't tell whether he was talking to his wife on the phone, or to me.

'Eric?' I say.

He wedges the phone between his shoulder and his ear, unreels a length of tape and tries to rip it with his teeth. The wind is strengthening again.

'Eric?' This is ridiculous. I should know this. I have my hand on the gearstick, wondering whether to rev the engine and go for it, watching the depth finder. The race is getting closer.

A herring gull is standing on top of the black-and-red one and the white circle seems very far over, almost concealed by a rhododendron bush. Does white mean stop, or foul ground, or mortal danger, or what? I stop looking at Eric and stare fixedly at the seagull on the post.

We're down to eight foot. Seven. Six, for God's sake, and I still have no idea what a white or a yellow pole means. I put the boat into neutral, which means it slows and then turns. It's the same feeling as yesterday, except that this time the water knows exactly where it wants to take us. It's disrupted and angry, tugging hard, trying to yank us towards the shore.

Eric looks up, sees what's going on, gesticulates. 'Bella! Pu ... t ... ttl ... !'

I can't look at him because I'm trying to keep control of this thing. Six and a half. Six. The skipper of this boat has entrusted me with 38 feet of fragile fibreglass and his only son, and I'm about to throw the lot on the rocks. I am shaking again. I push the boat back into gear and swing the wheel round. Too far round. Now the boat is side-on to the shore. Eric is yelling something at me, the wind's in my ear *and I can't fucking hear.*

Luke, who has been sitting on the back locker with his phone, looks up and says something. When I don't respond, he says it again. The third time, he puts it down and taps me on the shoulder.

'I'll do it,' he says, staring into my face, making his words big.

'No!' I say. I can't hand over the boat to a nine-year-old. 'Just tell me what the marks are!'

He looks over at the right bank.

'The white one! What's the white mark?'

He looks at it again, then back at me. Eric is advancing with the phone still in his hand, saying something. I can see the shapes his mouth is making, but not the sense.

Luke positions himself beside me and puts his hands

– his small-boy's hands – on the wheel. 'It's a sign.' He moves over. 'Look. I'll do it.'

Sure enough, as he pulls to the side, the boat swings round, just enough, not too much. He pushes the gear lever down a couple of centimetres so the sound of the motor tightens. The depth finder goes into freefall – six, then suddenly ten, then back to six again, then thirteen, then twenty. The boat is now heading in the right direction. At full revs, we round the tight corner through the buoys with inches to spare.

Eric has reached us. 'What happened?' I can see the anger in his eyes.

'It's fine, Dad,' says Luke. 'She didn't know what a road sign was.'

'Sorry,' I say. I don't look at him. Or Luke. I get up, walk down the deck to the bow, sit down, cry.

The humiliation I can stand, but the stupidity I can't. I shouldn't have come on this trip – all I did was endanger the others. I am angry with myself for being incompetent and I am angry with my ears for being unable to hear. It's such a simple thing, the act of hearing, so easy, so why can't I do it? I'm perfectly healthy, perfectly intelligent, perfectly capable of learning what needs to be learned, yet I seem to be too idiotic to steer a boat or hear a command. Somehow, I must have done this to myself. The deafness is a judgement, a physical sign of moral infirmity, proof I am somehow unsound.

At the head of the loch is calm water. The other boats swing on their moorings surrounded by the smell of blossom and wet grass. We pack and clean, unload sacks of rubbish and unclean clothing onto the pontoon, return to our cars. The goodbyes are perfunctory; all of us are keen to get away.

THAT NIGHT, AT HOME, I close my eyes. I see a version of myself lying on the ground. I'm curled up, not moving. I see another two or three versions of me standing above, kicking at the body below. The figures move round, top to bottom. Stand, balance, step back, consider, kick. Particularly the head and especially the ears. Then the heart. Most of all the heart. I lie on the ground, but I don't resist. The figures aren't angry. They're just neutral, inexorable. Almost scientific. Step back, consider, kick. Doing a good job. Nothing left unbroken.

2

Hearing

BY THE SUMMER OF 1998, it had become clear that there was something wrong with my hearing. It didn't happen suddenly – it wasn't like one week it was 20/20 and the next week it was down to 15/20 or 10/20 – but softly, so softly I almost wasn't aware of it happening; sound seemed to have stolen away. There was no pain, no sound of sound retreating, just the gradual understanding that something was less.

In January I'd been able to hear the traffic outside in the street. By March I could hear a few auditory exclamation marks – the bang of a door slamming, the blare of a horn – but not the noises that linked them. Noises which had been vivid seemed muffled while sentences which were once bordered by clear lines were now smoothed to a blur. Without the definition to speech – the sibilants, the corners and turns, the verbal signposts – I couldn't seem to find my way. Meetings became no more than a low seaside roar, and I kept connecting with the wrong end of a sentence. All the things which had once been so easy to navigate were now full of blunders.

For a while I did what any other sensible, evolved adult would do in the same situation – I ignored the problem. When that was no longer an option, I made an appointment with my GP, who referred me to the audiology department at St Mary's Hospital in London.

Steve the audiologist was in his early forties with pink

shoes and a tie with images of Scooby Doo on it. His skin had the faint luminescent sheen of someone who rarely saw daylight and for whom exhaustion is a basic state of being. He took a history and then directed me to a soundproofed room containing a pair of headphones before he took his place inside a separate booth filled with dials and machines.

He clicked a switch and his voice appeared in the headphones. 'Just nod when you can hear something.'

After a couple of seconds I could hear a low electronic tone, a single note. As I heard it, I nodded. The next tone was maybe five notes higher. I nodded again. And so it went, up the scale from bass to soprano. Across from me I could see Steve marking up his notes.

'Here,' he said when we'd finished, passing me a couple of sheets of paper on which were printed two sets of graphs, one for the right ear and one for the left. On each graph, the vertical side was marked from 0 to 100 with a black line at 50 in the centre. The horizontal side showed eight lines of equal thickness, one for each additional increase in volume. Steve had pencilled in a line linking the marks he had made showing the exact point at which the sound had become audible. Each of the lines swung up and down a bit, but none of them looked exactly 100 per cent.

I looked at the graphs and then back up at Steve.

'It's not great,' he said. 'I'm not sure exactly what's happened because the picture we're getting is contradictory, but it looks like those head injuries you told me about may have put pressure on your temporal lobe. You've pushed your skull down towards your spine and that's pulled all the muscles in your neck tight, which means you've ended up with less mobility in your neck than you should have. That means in turn that all the bones of the ear have been pulled around, and if those bones aren't able to move properly, then they're not conducting sound as they should do.'

'So,' I said, 'I've got a bottleneck.'

'Kind of.'

'And if the pressure could be relieved, will this get better?'

Steve looked at me. 'I don't think so,' he said. 'The pattern of hearing loss that you've got is inconsistent, but what I'm seeing suggests there's been some damage to the hair cells in the cochlea. So I doubt very much that the situation would improve even if you did straighten your neck out.'

'But if the bones could move again ...'

'The damage has probably been done.' On his desk was an unsexed human half-head, a model showing the side of a skull and the mechanism of the ear modelled in primary plastic. He picked it up and pointed at a red bit.

'Do you know anything about how hearing works?'

From school biology I knew that the ear had inner, middle and outer sections, but that was about it. 'No.'

Steve pointed at the portion of the model closest to the centre of the skull. 'The trouble is that so much of hearing takes place in a part of the body that we can't access. If you break your leg, we can X-ray. If you get cancer, we can do a biopsy or a scan. In most situations, we can get an actual physical picture of what's going on inside you. But with the ear, we can't. We can do an MRI but even that won't give us that much information. If you had punctured one of your eardrums, we'd be able to look at the damage from outside because it's part of the outer ear, but we can't see inside the inner ear because the inner ear is way down deep inside your skull and the cochlea is a tiny enclosed spiral within that space.'

He set the model down on the desk in front of me. 'All we can do instead is to look at the clues provided in your audiogram and, from the pattern of it, make an educated guess about what's going on. It's usually a pretty accurate guess but it's still a guess because we can't physically see in there. And, generally speaking, for exactly the same reason, once anything within the inner ear is damaged, it stays damaged. In most cases, we can't do anything about it because the inner ear is so very inner that surgery on it wouldn't be ear surgery, it would be brain surgery.'

He paused to give me time to digest this. 'The good thing is, your high-frequency hearing is still reasonable.'

I looked at him, blank.

'OK,' he said. 'Hearing loss isn't just like going from total hearing to total silence. It's not something you have and then you don't. Or not normally anyway. Everyone hears across a range of frequencies from high to low. When it's at its best – usually when you're young – the human ear can pick up everything from twenty hertz to twenty kilohertz. But that's unusual. What happens normally is that people get older or they're idiots and they go to festivals and stand too close to the speakers and their hearing across certain frequencies starts to wear out. But it doesn't wear out evenly across all frequencies.'

He turned the plastic model round to face him and pointed at a tiny shell-shaped spiral deep within the inner ear. 'Do you know anything about the cochlea?'

I shook my head.

He kept his finger against it for a second. 'The cochlea is minute. It's no more than the size of a sunflower seed, but it's lined with thousands of cells called stereocilia. Because they stick up like the tufts of a rug and move with the vibrations of sound, they're known as hair cells. And because it's a spiral with the high-frequency hair cells at the front end and the low ones in the centre, that means that even if you're hearing a low-frequency sound, it has to pass over the high-frequency hair cells before it gets to the low ones.'

I nodded. I wasn't taking in much, but it was good to show willing.

'And because the high-frequency cells are getting so much more traffic, they tend to wear out more quickly. Think of it like a rug. Like any rug, it gets worn more in the parts which get heavily used than the ones that don't. So the high-frequency goes first, then the mid, then the low.'

He pointed back at the top two graphs of the audiogram. 'You don't seem to have lost that much of your high frequencies, but if you were to average out the loss across all areas, I'd say you've lost about forty to fifty per cent of normal hearing.' He paused, still looking at the page. 'I'm surprised that you've put up with this for as long as you have.'

'But,' I said, 'I'm twenty-seven. Why am I losing my hearing?'

'But,' he smiled, 'you keep hitting your head on things.'

I smiled back, perversely cheerful.

'If you were a bird,' he said, 'it would be different. In humans, the hair cells don't regenerate. You get one lot, and that's it. But in birds, if those cells are destroyed, they can grow back. Because if birds can't hear other birds they can't find a mate, and if they can't mate then the species can't survive. Plus of course it always helps to hear a predator sneaking up on you. But even if every hair cell is destroyed, they grow them again. They can hear again within a month.'

I wasn't interested in birds. I was interested in whether this was going to get better or not, and now I knew it wasn't. I'd never heard of a cure for deafness. There were no cures for deafness.

Steve looked at me and for a moment I saw a whole other conversation open up in his eyes, a conversation in which there was compassion and a kind of acknowledgement. Then he glanced at his screen and pushed his chair back.

'Look,' he said, 'I know it's a lot of information to take in. But for the moment, what we'll do is show these results to the doctor. And then we'll fit you up for some hearing aids.'

BACK IN 1990 I had gone together with a group of university friends on a skiing trip. Somehow one of us had managed to pull off a cheap end-of-season deal in one of the big French resorts, an all-inclusive thing with a self-catering apartment plus a coach trip at either end. It was booked for right at the end of March and, whatever our level of skiing expertise, all of us were looking forward to it.

It wasn't until the coach pulled up opposite the ski lifts that we realised there was no actual snow. Nothing. Not a flake. Cheap, in this instance, meant dustbowl. Higher up the mountain, the ground was bare and flat, the soft brown of new hide. Farther down, the grass had already begun to return, shading what would have been the beginner slopes in fresh green. Outside some of the larger hotels the snowdrops had already come and gone while daffodils bloomed by the roadside. For a moment we sat back, enjoying the sound of meltwater rebounding down the gullies.

'Oh well,' said Clare beside me. 'Never mind. How about rollerblading?'

Beside us, an Italian coach pulled up. Through the windows, we watched them react as we had: dismay, laughter, animated chattering.

There was no point in remaining where we were, so we got out and unloaded our bags.

'*Pas de neige?*' I said to the woman in charge of keys for the apartment, pointing over my shoulder. '*Où est la neige?*'

The woman smiled. '*Pas de problème!*' She gestured towards a kind of thin white B-road twisting down the mountain beside the main ski lift. It glittered in the sun, not the powder-blue shadow of real snow, but the hard refraction of ice. '*Neige artificielle! Pour tous!*'

'*Oh, oui, d'accord!*' we agreed politely, '*Absolument!*'

The next morning, having kitted ourselves out with the necessary boots and skis, we set off. Only a few of the lower lifts were still open and the remaining runs had a thin trail of compacted permafrost covered with a dusting of artificial snow replenished nightly by the snow cannons. The pisteurs had been up and down the runs with their ploughs, ridging the ice in brittle lines. Farther up, a few blobs of unmelted white helped to give a feeling of what the place might look like at the moments when it looked like it was supposed to look.

Despite this, the runs which did remain open were full. Children hurled lumps of mud at each other instead of snowballs, and parents sat sunbathing in the cafés. For the first couple of days we got used to the impressive braking effects of sliding from ice to earth, stumping up and down from the cafés in T-shirts and sun cream.

On the third day we all decided on a trip farther up the mountain. In the cable car, hikers in boots and anoraks grinned at us or unfolded maps showing the network of walkers' footpaths. Even up here, right at the top, only about half the runs were open. In every direction the great jagged panorama of the Alps lay before us. The higher mountains were still tipped with white but wherever we looked, the murk of the snowless valleys crept almost to the summits. We split off into different groups according to ability. Three of us wanted to see whether there was any off-piste worth skiing at all, while there was one long red I wanted to try. The others set off, skis rattling. I looped the straps of the ski poles round my wrists, turned on my Walkman and headed off.

The weather was bright and clear, and the contrast between the bits where the sun reflected off the ice and the black behind the rocks was so stark it was sometimes difficult to see even with sunglasses on. Up here, the problem was less in sliding from ice to stone – which was at least visible – but from those few patches of snow back to ice again. Either way, the result was fast and unpredictable. I could hear the different textures beneath my skis – the bony clatter of compaction and the punctured hiss of slush as I twisted downwards, skiing more by feel than by sight. Beyond the sounds of my own skis, I could hear birdsong – not just one bird, but hundreds, thousands, a great spring chorus all singing for their lives in the turn of the Alpine seasons. I wasn't wearing a helmet so I could feel the warmth of the sun on my head and

I was beginning to hit a nice easy rhythm when I hit one of the invisible patches of ice, went back on my skis, spun out of control and slammed head first into a rock.

For a couple of moments, I must have blacked out. Then I came to, lying beside the rock with the skis and poles scattered somewhere behind me, levered myself up and ran a hand over my forehead. It hurt, and there was quite a lot of blood. As I levered myself to my feet and began groping around for the lost poles, two skiers appeared behind me. They looked at me, then at the rock, then at the blood on the ice. One of them helped me retrieve my belongings and the other said something urgent in German, repeating the same phrase several times before they looked back at me one more time and then skiied away fast down the mountain. I wrapped my scarf tightly around my forehead, stamped my boots back into my skis and set off very slowly in the same direction as them.

It can't have taken that long to get to the end of the run, though it felt like many things happened in that time. It wasn't so much that I was having difficulty with the ice, it was more that something seemed to have happened to my conception of three-dimensional time-space. During the sixties those in search of the ultimate mind-altering experience were fond of drilling holes in their foreheads, a practice known as trepanning. The theory was that if you opened up the route to the frontal lobe of the brain – the bit then thought to filter consciousness – then you removed those filters, thus awarding yourself a VIP pass to the seventh dimension. By smacking myself against a rock, I'd just done the same thing to myself for free. So on the way down the mountain I was experiencing two completely separate versions of reality, both equally persuasive and equally complete. In one, I was inching my way along the ice, half blinded by my own blood and staring at the few inches of ground in front of me. In the other, I was back at home having a vigorous argument with a friend about cooking. The argument wasn't a dream or an hallucination, it was real – as real as the mountain. Some atavistic part of me knew it was vital I kept a grip on one of those realities, though I couldn't always seem to figure out which one.

I don't know how long it took for the tussle between the ice and the friend to resolve itself in favour of the ice, but at some point later I arrived at the base of the run and the little wooden cabin beside the ski lift. As I appeared, three Ecole de Ski first-aiders were adjusting their goggles and making last-minute checks to the blood-wagon sled. I stood in the doorway.

'*Je pense que j'ai besoin d'aide,*' I said. '*J'ai mal de tête.*'

One of the guides looked at me kindly. '*Allez,*' he said, shepherding me into the hut. '*Asseyez-vous.*' I sat down and unwrapped the scarf. He looked at me and smiled. 'I think,' he said in English, 'that looks like more than a headache.'

'I'm fine,' I said, and started to cry.

Down at the bottom of the mountain, the local doctor X-rayed me before running a neat seam of prickly stitches from the top of my forehead into my hairline.

'Go home,' he said. 'Really. Forget trying to ski. Go home.'

When we met again at the apartment, the others did their best to persuade me to do the same, though I had no intention of going home. I was twenty and therefore immortal, and as far as I was concerned there was nothing about a major head injury that couldn't be solved with Nurofen and paint-stripper wine. I think I did get an upgrade from the floor to the sofa, but when my friends saw I was determined to stay, we all just carried on as normal.

At the end of the week we took the coach back to London. I had found a cap to cover the worst of the mess on my forehead but my face had swollen up and the area around my eye sockets had begun to bruise. It hadn't occurred to me to tell anyone at home, so when my father opened the door he saw a stranger, a rainbow-hued B-movie monster.

'It's interesting,' he said once he'd got over the shock. 'Almost Cubist.'

'You look better like that,' said my youngest sister, round-eyed. 'Will you always be purple?'

SEVEN YEARS LATER, when the scar on my forehead had faded to a discreet white line, the ice and I met again. As I drove out from Edinburgh on a frosty March night, the car hit a patch of black ice, skidded, hit the hedge on the other side of the road and flipped over.

I came to rest upside down, my head separated from the road by a thin strip of metal. The driver's window had shattered in the impact, and I could hear diesel spattering out onto the road. I wriggled out of the window and stood for a while on the verge, shivering with shock. A haulage truck travelling in the opposite direction stopped and the driver rang the emergency services.

Not long before I'd seen the David Lynch film *Wild at Heart* in which the two main characters come across a recent car accident in the Nevada desert. One of the victims looks OK – she's standing, she's worrying about her purse – but shortly afterwards she dies in Nicolas Cage's arms. Finding myself by the side of the road on a dark night after a car accident, I thought about the film again. Maybe I looked all right on the outside, but what if I was like the girl in the film, just seconds away from collapsing into a haulier's front cab? I didn't want that to happen, so until the ambulance arrived I crouched by the trees and recited my name and address over and over in case I'd left something important – memory, identity – by the side of the A702.

It was shortly after the accident that I had started to go deaf. Abruptly, life seemed to have become one long cliché: pushing up the volume on the TV or the stereo, needing people to repeat things, missing the phone or the sound of my boyfriend Euan's key in the lock. I had to be up close to hear the hiss of unlit gas on the stove. Euan complained that I'd stopped laughing at his jokes and even when I did it was three minutes later.

It would probably be good to say that it was the decline of children's voices or the vanishing of Euan's laugh that finally forced me to go and do something about it, but to be honest, it wasn't that – it was the fear of being bad at my job. At the time I was working as a freelance journalist and had just started on my first book, an account of the Stevenson family of engineers who built all the Scottish lighthouses and then produced Robert Louis Stevenson. Much of my time was spent in archives going through family papers, but my researches also took me out to the darkest corners of Scotland to interview the last generation of lighthouse keepers.

Keepers, I soon discovered, fell into two distinct categories: unstoppable, or silent. Either all I had to do was switch on the tape recorder and let it flow or I was there for hours, tweezering out each individual word. But when I transcribed the interviews afterwards I noticed how often I had misheard or leapt in at the wrong moment. Admittedly some of the interviews had been conducted outside and lighthouses are by nature high and windy places. But even so, there seemed to be a lot of gaps. All the grace of a conversation had gone and now there were only a series of jolting observations. I could hear how hard I was trying but I could hear how hard the keepers were trying too – the note of bemusement in their voice when I failed to pick up on something they'd said or came back with something completely unrelated. They were all polite, but behind that politeness was a distance, a gap which I kept trying to cross but which always slid farther away.

It was frustrating, sitting there at my desk, turning the volume up on my own mistakes. Every time I listened to one of the tapes I could hear – at top volume, and as many times as I chose to repeat it – just exactly how many times I hadn't been able to hear.

3

Aid

UNTIL I'D TURNED THE CAR upside down I had been just like everyone else – I accepted the ordinary miracle of my senses and I expected them to get on with the job. For the past twenty-eight years I'd lived my life with perfect hearing and all the gifts that came with it: speech, communication, language, a link to other people, the freedom not to even think about the efficient functioning of my own body.

In the weeks after my appointment with Steve, I flicked through medical guides or skimmed web pages. Somewhere between the graphics I was aware that I wasn't searching for a cold description of process but a line of hope, a few sentences shining out from the page: 'Of course, in your case, you'll definitely get better from this. In your case, we're pretty sure it's just wax.' But the medical books didn't say that. What they told me instead was a bit more about how we hear and sometimes why, though they never really said anything about what happens to us when we stop.

As Steve had pointed out, part of the genius of hearing is that so much of it takes place so deep inside our heads. The bit of the ear that we can actually see is the doorway to a line of secret chambers tucked deep into the skull. Each of those chambers is linked to the next in a chain of connection reaching all the way from the eardrum to the brain.

All sound – low or high, loud or faint – travels in waves. It's relatively rare to be able to experience it as one, but it's possible, particularly at low frequencies. The opening notes of Bach's *Toccata and Fugue*, a truck rumbling past, dubstep – they all resonate through our bones. At higher frequencies those resonances appear to become more a sensory experience than a conductive one, but they're still made of waves. It doesn't matter whether a sound is high, low or ultrasonic, it's there because something has rippled the shape of the air in the same way as a boat's prow makes a bow-wave through water – small ripples for high-pitch frequencies, big for low. Every noise, whether it's the song of a skylark or the crack of a rifle, is made of the same three things: air, pressure, time. And while frequencies are measured in kilohertz (kHz), their volume is usually expressed as decibel Sound Pressure Level (or dBs SPL). In other words, the notes up and down a piano would be measured in kHz, but the loudness at which they were played would be dBs SPL.

When the ripples of sound reach the ear, they're gathered and channelled by the hard cartilage of the pinna (or auricle) into and down the ear canal to the eardrum. As its name implies, the eardrum is a taut membrane separating the outer ear from the middle ear which looks in cross-section a bit like a pea seedling. Like a musical drum, it vibrates as it's struck by the incoming sound waves.

Behind it is an air-filled space containing the three smallest bones in the human body: the malleus (hammer), incus (anvil) and stapes (stirrup). Each of those bones has a separate function. The malleus is attached to the eardrum, the stapes connects to the inner ear, and the incus links the two. When a sound wave reaches the eardrum, each of those bones (or ossicles) vibrates in turn, transmitting and concentrating the sound as it connects to the inner ear and the cochlea. That little snail-shell contains not just the 2,700 individual hair cells (cilia) which Steve had been describing, but three separate chambers filled with fluid. Vibrations in the fluid stimulates the hair cells, and the cells release chemicals which in turn activate auditory nerve fibres to transmit information about the type and quality of that vibration through to the temporal lobe in the brain.

All the processes involved in the outer and middle ears are mechanical. In each stage, something physical happens – the pinna gathers, the eardrum vibrates, the ossicles conduct. It isn't until the sound reaches the inner ear that the representation of that original ripple ceases to be a physical process and starts to become a sensory one. All that's happened in the outer and middle ear is that different wave patterns have been used to generate variations in pressure within the fluids of the inner ear. It is the hair cells that transfer them from a purely mechanical energy into electrical signals, turning them from a form which the brain cannot comprehend into a form that it can. If you think of sound as cryptography, the outer and middle ears are the satellite dishes receiving the signals, but the temporal lobe and the auditory nerve are the parts doing the deciphering. In other words, sound is received and processed in two parts. The ear receives, the brain processes – a physical Enigma code.

Because there are so many parts to hearing, there are an equal number of parts to hearing loss. Generally speaking, they can be split into three. There is conductive loss in which there's been a failure of some part of those initial mechanical processes such as puncturing of the eardrum or damage to the ossicles. Or there's sensorineural loss in which the nerves or the hair cells have been damaged. Partly because that damage is often caused either by ageing or external environmental factors, it's a much more widespread form of hearing loss than the conductive type.

Any very loud noise may eventually damage or disrupt the stereocilia and lead to the death of hair cells, whether that's an iPod always on full volume or listening to aircraft taking off at close proximity. Sometimes that damage results in hearing loss and sometimes it results in tinnitus, where true sound is replaced by sounds the brain itself has made. Some people hear hissing or fizzing or clicking, or a steady background noise like an internal fridge. Tinnitus can be loud or soft, constant or intermittent, but of all the different kinds of hearing disorders it's often considered the worst because it covers over the sounds that people want to hear with a drizzle of sound that they don't. And finally, there's a mixture of both conductive and sensorineural loss, which is what the composer Ludwig van Beethoven had.

IN THE EVENT it took several months for the correctly fitted analogue aids to arrive. After a couple of false starts, I returned to Steve's office in St Mary's in the spring of 1999. As I sat down beside his desk, he passed across a small black hexagonal box in which were two hearing aids designed to fit down the canal to the eardrum. 'Property of the NHS,' said the lettering on the box.

I lifted one of the aids out and looked at it. Steve had said they were going to be discreet in-the-ear aids, so I'd had an image of something almost invisible to anyone looking at me face-on. But these things seemed enormous, great blocky lumps of plastic in the same flat grey-beige shade as a hernia gusset. At the inner end was a little hole for the sound to reach the eardrum and at the outer end was a square lump with a slot for a battery, a volume control and another, smaller hole to let the sound in. Compared to the abstract splendour of an eighteenth-century ear trumpet these were miracles of miniaturisation, but to me they looked as if you could have fitted long wave, short wave and FM onto them with ease.

'Try them,' said Steve encouragingly.

I slotted one into my right ear. Then I took it out again.

'You need to wear these all the time for them to work.'

'Why?'

He looked at me. 'How much of what I told you last time did you take in?'

'Some.'

'OK. Well. Just take it from me that in order to get the benefit from these, your brain needs to get used to them, and your brain will only get used to them if you wear them all the time. These things are only miniature microphones and loudspeakers, but they're the best we can do. There are digital aids coming onto the market and hopefully some day soon the price will come down enough for the NHS to be able to offer them. But at the moment this is what's available. If you use them right, they'll make life much easier for you, but it will take about four months of full-time use for the brain to adjust to processing sound through the aids.'

He put the aids back into their box and pushed it towards me. 'So,' he said. I suspected he knew exactly what I was thinking. 'You need to actually put these things in.'

Can I customise them? Is my hair long enough to cover them?

'Go away,' said Steve, exhaling. 'Try them, get used to them. But for God's sake just use them.' He turned back to his screen. 'Can you talk to reception about an appointment in six months' time? Say, mid-May?'

AS IN SO MANY other things, I was – I am – astoundingly fortunate. I was born at the tail end of the twentieth century, a time in which the science of otology and acoustics has been racing along at the same rate of advance as neuroscience or genomics. I had the benefit of tiny, in-the-ear hearing aids, of science and of the drive to produce something that really works. Those born before me did not.

Somewhere deep in the recesses of Blythe House in London, thousands of objects from Henry Wellcome's medical collections are interred, including a selection of early aids ranging in origin from the late seventeenth century to the early twentieth. Among them are trumpets of every conceivable shape and material from the practical to the hilarious. There are elegant double shell shapes ergo-nomically sculpted to catch sound in tiny tortoiseshell side buckets next to primitive tubas shrunk down to a tenth of normal size. There's a small Victorian bicycle horn next to what appears to be a series of miniature policemen's helmets with teapot spouts attached. There's a small chemical retort made of copper in a drawer also containing what must surely be the dispossessed end of a clarinet. There's something which, when stood on its end, looks exactly like an old kettle. There are two big shallow circles made of black papier mâché designed to sit behind the wearer's ears as ready-to-wear sat-ellite dishes. There are things in wood, cloth, tortoiseshell, shellac, papier mâché, brass, copper, Bakelite and steel, some of which would have conducted sound beautifully and some of which would probably have swallowed every last particle. There's even a Victorian mourning-trumpet, its opening so clogged with black lace that it could only have been used by someone who had absolutely no intention of hearing any-thing at all.

Looking at the variety in all these different drawers, two things become clear. One, that for as long as ear trumpets remained the main way of amplifying sound, no one ever reached agreement about the most efficient form of design. And two, anyone venturing out with one of these in public needed considerable resilience of character.

In Bonn, the Beethoven Haus has four of the ear trum-pets once belonging to the composer. All are made of copper, and all are elegant variations on a horn shape – long taper-ing stalks flaring down to an open end. Three of them are shaped into the bell of a conventional trumpet and one has a little circular box with a perforated speaking surface like a primitive telephone receiver. Two were designed to be worn with headbands, leaving the hands free but present-ing a challenging look even without the composer's freestyle approach to personal grooming. Each design has a narrow end designed to fit down the ear canal and a wide receptacle designed both to receive sound and to reflect it inwards. The irony is, of course, that all four of them look less like medical aids than musical instruments. Any wind instrument works by channelling air pressure waves into a particular shape so that those waves then resonate with a note of a particular pitch. It almost certainly wasn't lost on Beethoven that he had reached – and then passed – the point at which he now required a trumpet in order to hear a flute.

Beethoven's relationship with the trumpets was as turbulent as most other things in his life. In 1812 he bought the first of several 'hearing devices' from the Viennese inventor Johann Mälzel. Over time he experimented with different shapes, searching for the optimum amplification, and by 1815 had concluded that 'One should have different ones for music, speech, and also for halls of various sizes'. Some of his visitors reported that he used the devices for conversation, and some that they just had to yell. By 1820 he'd stopped using all of them, partly because his hearing was now so bad that they no longer made a difference, and partly because (judging by the number of dents) he often got so frustrated he yanked the trumpets off and hurled them across the room. When he did finally give them up, he accused them – and Mälzel – of further damaging his hearing, and wrote dismissively that 'They were not of any real use to me'.

Compared to the first generation of electronic aids which came with more wiring than the Apollo space programme, even the most basic modern digital aid is a technical and technological marvel. There are still plenty of people around who remember the aids of the 1980s – solid chunks of plastic worn behind the ear connected with a wire to a transistor worn on a chest harness. Even if the wearer looked perfectly intelligent beforehand, just putting on one of the boxes appeared to drop their IQ by a good 70 points, and as a great big mid-chest bullseye for bullies, they could scarcely have been bettered. Even once the boxes disappeared and all that was left was the first generation of behind-the-ear aids, the best result possible was still somewhere between Unfortunate and Low-Key Remedial.

It wasn't until the late twentieth century that hearing aids began to shrink to the point where they could be worn relatively unobtrusively. As the aids got smaller, the difficulty for manufacturers was that they had to fit a great deal of complex technology into a very small object. Effectively they have to cram an entire mixing desk into a strange-shaped space which would, at most, only be a few millimetres long. Both analogue and digital aids have to allow sound to enter effectively and then to pass that amplifed sound through to the eardrum, while digital hearing aids also need the technology to discriminate between different frequencies. Unlike the analogue aids, on which only the volume can be controlled, digital aids can be tuned by an audiologist so, for instance, they give greater amplification on high and mid frequencies, but less on low. By doing so, they help the human brain to discriminate between important noises (voices) and unimportant ones (traffic). Which, you might say, is what true hearing does. But that's not quite right. True hearing edits all the time. Every second of every day it judges and discards, picking through what it understands to be significant and ignoring everything else. So sticking a small pair of amps in your ears might help a lot with the volume of what you hear, but it'll do absolutely nothing for the sense.

Meanwhile, cochlea implants are becoming an increasingly significant form of treatment for conditions of the inner ear. An implant works by bypassing the damaged hair cells completely and passing signals directly to the auditory cortex. At the moment they're only given to those with severe or profound hearing loss caused by damage or deformation of the cochlea, though the age of implantees drops every year – at the time of writing, the youngest recipient is a baby of three and a half months. But, as with hearing aids, there's always a time-lag while the brain learns to translate these strange new soundings into something containing sense and fluency. Which means that, broadly speaking, the younger the recipient, the greater the chance their brain will learn to process these new signals as meaningful – and conversely that when older, deafened, implantees hear voices or music for the first time in many years, it may well sound horrendous.

Hearing isn't a single-cell process – it takes place in many parts of the brain, which means its effects are registered in all parts of the body. But its full complexity only really becomes apparent when placed beside science's artificial alternatives. It's like hands, or kidneys. The greatest minds in medical invention have been working on a prosthetic hand for generations and though they've undoubtedly reached a point of great sophistication, the results don't even come close to the real thing. As for twist-grips, so for cilia – all our brilliance can only produce a rough imitation, a golem to the sheer evolutionary beauty of true hearing.

IT WAS MAKING that May appointment which finally changed things. Though by then it had been nearly eighteen months since I'd started losing my hearing, I'd somehow managed to delude myself that this was something temporary. A long-running totally asymptomatic case of flu. Uncleared air-pressure build-up –anything that allowed for the possibility of waking up one morning with the wool pulled out of my ears. And, though I was aware that the NHS didn't traditionally give away expensive items of customised hardware to audiology outpatients, I'd even managed to convince myself that they too were available purely on a trial basis.

Until that moment I'd taken all Steve's information in but I hadn't really absorbed it. I'd understood it, but I hadn't comprehended it. Now, finally, it dropped into me – all of me. If Steve was suggesting appointments several months in advance, it meant that he knew that I would still be deaf in six months' time. Or a year. Or quite a lot more than a year.

I didn't wear the aids. Or rather, I didn't wear the aids as they should have been worn. I didn't like them, and I found them uncomfortable. Instead, I bought a couple of small pots of brightly coloured nail varnish and customised them, painting out the gusset-coloured plastic and adding a bit of glitter. When I put the new sparkly aids in they felt better. I wore them occasionally, taking them in and out according to circumstance. If I was in a room with one or two other people the aids could be helpful but I soon found they amplified all sound, not just the ones I wanted to hear. Outside in the street the volume of traffic noise overbore the voice of the person I was talking to. Inside, the sound of the washing machine drowned out conversation. On the phone they were screechy with feedback and got in the way. I couldn't seem to get the volume right.

When I returned to St Mary's in May, Steve was still wearing the pink shoes and the Scooby Doo tie, and looked if anything marginally more exhausted than he had six months ago.

When I produced the coloured hearing aids, he laughed. 'I approve. Anything that makes you happier about wearing them has got to be a good thing.'

It was at that point that I finally plucked up the courage to ask the only question that seemed relevant. 'Is it going to get worse?'

Steve looked at me. 'Probably,' he said. 'It will probably go so slowly you scarcely notice. In two years, who knows? You could be a psychological jellyfish. All your social skills could be completely eroded. You'd still be hanging in there, but you wouldn't really be able to get by in any kind of normal human interaction.'

'Right,' I said.

'Or,' he said, putting the aids back in their box and handing it back, 'you could wear these.'

4

Loss

IT DID NOT TAKE ME LONG to adjust to being deaf. Or rather, it did not take me long to realise that I really didn't want to be deaf, and that – faced with a choice over whether to go gracefully or to yank the building down around my fading ears – I was going to give it everything I'd got. Metaphorically speaking.

On the outside I did my best to sound as if everything was fine. But inside I was hurling myself around the bars of my self-made cell from dawn until dusk, trying to claw my way towards the invisible adversary who I believed had somehow made me deaf in the first place. I knew one thing: I wasn't going to be graceful about this. I wasn't going to go quietly, and I had absolutely no intention of being even slightly well adjusted about it. If this was a kind of bereavement and bereavement was supposed to have four stages, then forget all that nut-roast stuff about submission and acceptance. I planned to stick right here on fear and denial with maybe a bit of cosmic plea-bargaining thrown in for good measure. I could be sane and proportionate about this, or I could turn a manageable development into a full-blown existential calamity. If there was a hard way to do this and an easy way, then I was going to march straight towards the door marked bloody impossible.

Of course, at the time, none of this felt like a choice. Something had happened and I'd reacted to it. That reaction seemed inevitable; I didn't choose to feel terrible about going deaf, I just did. It's only now that I understand there were other ways of reacting. But when it actually happened, I couldn't figure out a way to be graceful when all I felt was ashamed.

At the time, deaf meant two things to me: stupid, and old. As far as I was concerned deafness worked on people like social cement – it slowed them up, glueing them to their last conversational footprint, stranding them at whatever point they'd misheard. It didn't matter what they'd done or who they were, deafness levelled everyone down. Before, they might have been heroes or comedians; now, they were always a beat behind, a joke away from the punchline, two repeats over the moment. They were the ones who trailed along, waiting for a helping hand.

And old people. My impression was that old people were always deaf. As Steve had said, deafness was what happened when bits of you started wearing out. It signified the moment at which people started discarding all the things – sight, mobility, hearing – that once gave them a connection to the world. It speeded the pace at which they turned away from other people or started believing that silent past was better than cacophonous present. Once hearing began to go, it was like watching the cables that tied someone to life snapping one by one. To the rest of the world, deafness was what hap- pened at the end, not the beginning, and give or take the odd issue with dentures or incontinence I couldn't actually think of a condition that seemed more ridiculous. I was now 28, but that was it. *Hearing loss. Deaf. Pathetic.* So pathetic that when I told people about it, most of them just laughed.

'I'm going deaf,' I said to Euan.

'What?' he said.

'I'm going deaf.'

'Pardon?'

'Not funny. Not funny at all.'

'*Quite* funny,' he said. 'To me, anyway.'

As for the rest of the world, the first thing was to mini-mise the scale of the problem. OK, so I could hear something through these new hearing aids, but the really important thing was to pretend I could hear everything.

There were various different ways of doing this. First and most obvious was to watch the speaker's lips. All of us, whether we know we're doing it or not, lip-read. We don't necessarily all lip-read well, but we're aware of following the shapes that someone's mouth makes. On old movies or cheap film footage, the voice track occasionally slips out of sync with the speaker. Even if it's out by only a fraction of a second we can see it instantly and feel the effect to be wrong. So I grew used to watching people's faces, following their movements with what seemed like an almost indecent level of attention. At the time I was quite short sighted and because I was often putting on and taking off my glasses, it became very clear very quickly that I was at least 40 per cent deafer without them on. Without them, I couldn't see detail – the exact shape of certain words, the tiny shifts in emotional colour.

With my fellow human beings, my best option was to second-guess what was being said. If I was in a work meeting, then hopefully the room was quiet and the discussion would follow certain defined topics. If I was talking to a friend, then I could ask them about partners or children, which would at least narrow down their responses to that particular subject area. And if I really had no idea what someone was talking about, then I'd do my best to pull a full sentence's sense out of a single word, thus making every conversation a thrilling game of neurological hide-and-seek. If I thought I'd heard the word 'football', for instance, I was on fairly safe territory. All I would need to do was say something like 'So how did the game go?' or 'Haven't Chelsea got a new manager?' and then sit back, safe in the knowledge that whoever I was with would then talk animatedly for at least four minutes with no further input from me. But if all I'd heard out of a sentence was 'clear' or 'right', or one of the many other words in the English language with multiple possible meanings, what could I do? Did 'right' mean 'turn right' or 'write', or 'right versus wrong'? Was it better to make a stab at a reply, or to ask them to repeat? Could I come up with a plausible on-hold response, or was the speaker expecting an answer? Generally, my solution was a journalistic one: just keep asking questions.

The other thing was to push back as hard as I could against the effects. If deafness was supposed to be isolating, then I'd socialise twice as much. If it was designed to curtail work, then I'd work at double my usual rate. If it was going to stop me from doing interviews, then I'd do twice as many for three times as long. If it was supposed to tip me into myself, then I'd be five times as outgoing. I'd stop watching TV and go to gigs. I'd eat ten times the life I'd consumed before.

As for the shame, that was simple. I dealt with that by not talking about it. To anyone. At all. Ever.

Just before the first appointment at St Mary's, I'd gone for supper with an old friend and told him I thought I was losing my hearing.

'You're not deaf!' he said. 'You just don't listen!'

He had a point, and that point had struck home painfully. Maybe I had used my hearing wrong. And maybe it was my fault that it had been taken away. If I hadn't listened properly, then I deserved to lose the use of my ears. If I'd ignored friends or family when they were in trouble or failed to hear what they were telling me, then this was just the obvious result. If I had behaved like an island, then why the hell should it be a surprise when I became one?

AS IT HAPPENED, it turned out to be a very overcrowded island. Though I didn't realise it at the time deafness is a very common problem, as is not talking about deafness. Absolute figures are hard to come by because so much is not reported, but according to Action on Hearing Loss (AoHL, formerly the Royal National Institute for the Deaf), there are over eleven million people in Britain with some form of hearing loss – nearly one in six of us. On top of that, there's a relatively small number of deafened individuals who learned speech young but who became fully deaf later, and who may communicate both orally and with sign. Then there are the 150,000 deafened adults – those who were born at least partially hearing but who developed a severe or profound loss later in life. In the vast majority of cases, that loss is age related. Action on Hearing Loss estimates that 41.7 per cent of those over 50 have some form of hearing loss, and 71.7 per cent of those over 70 do. I was one of the 2 per cent who had lost their hearing young.

That one big hearing-loss tribe contains several smaller ones, though crucially, that 11 million does not contain those born severely or profoundly deaf, unable to hear anything below 70 decibels (in the case of severe) to 90 dBs (for profound), and often known as the prelingually deaf. Current estimates put the numbers of those for whom sign language is the main mode of communication at around 60,000. There are of course as many shades of grey as there are individuals, but if you wanted to simplify things, you could say that there are two categories: the Deaf – the 60,000 – and the deafened – the 11 million.

And most of those 11 million are no better adjusted about losing their hearing than I had been. Admittedly, my hearing degenerated quicker than most, but even with a relatively sedate decline, AoHL estimates that it takes an average of ten years for someone to actually acknowledge the problem. Ten years before they go to the doctor, get themselves tested or find some kind of remedy which works for them – a whole decade spent in reverse.

In part, that denial is because the process is slow. Unless the deafness has been caused by some form of trauma – an event which damages or breaks the eardrum, for instance – most people lose their hearing painlessly. It takes itself away a molecule at a time, fading out like the last contrails of a song, and its going probably won't even be noticeable. Its absence becomes present only in small markers and flag-days. But the denial is also because of that stigma. Turns out old people don't want to be deaf either. They too see it as a marker and find it as upsetting as I had done. For them it is also a rite of passage, and they don't much fancy the implications. Given that daft question about whether you'd rather lose your sight or your hearing most people choose hearing, but when it actually happens it's not just the loss of that sense that people – of whatever age – have to address, it's also the sadness of having lost it. Sight gives you the world, but hearing gives you other people. It gives you your capacity to interact, to use the gift of language and contact, to be heard and understood in the world. Take it away and you don't just remove the simple pleasure of sound, you remove your route through to humanity. It changes the way you work and the way your employer thinks of you. It changes your social life and your cultural interests. Above all, it threatens to change your relationships.

Which is probably why there are so many mental health issues associated with deafness. The incidence of depression is four times higher among those with hearing loss than within the population as a whole, and among the deafened (those with acquired profound hearing loss later in life) it's five times higher. There's also a much higher incidence of drug and alcohol abuse among the Deaf, of personality and behavioural disorders, and in some cases of something akin to post-traumatic stress disorder. And, as I discovered later once I started talking to those in the music and military worlds, it isn't just the victims themselves who are affected.

The incidence of depression among the hearing partners of deafened individuals is four times higher than normal, all of which is exacerbated by that ten-year delay in seeking help.

AS STEVE THE AUDIOLOGIST had noted, part of the problem with losing my hearing was that I knew nothing about hearing, so therefore I knew nothing about deafness either. What was I? Was I me, normal, except with hearing aids? Was I someone who was going deaf, or was I deafened, or was I just hard of hearing? Was I looking at a totally silent future, or did I just have this thing called hearing loss, a condition apparently defined entirely by absence?

The only thing I thought I knew about deafness was that there was a distinction between those who had been born deaf and those who had lost their hearing. Those who had been born deaf were, I understood, a proud and distinct community with their own history and politics. Once in a while I'd come across two or three people signing in the street and marvel at the ballet their hands were making. I knew that there were different types of sign language – British, American, French – and that deaf people had a whole network of clubs and working lives completely distinct from the hearing world. I'd always been intrigued by pubs where all conversations were conducted completely in sign.

But the proper deaf-since-birth world seemed a distant elite with no more relevance to me now than the blind-since-birth. I'd had my hearing and I'd mislaid it somewhere, so neither sympathy nor support was due. All those people who, like me, had possessed full hearing and were now losing it – they weren't proud or distinct. They didn't have their own culture or community, they were just a lump of people walking around with their hands cupping their ears. So it genuinely never occurred to me to look for help or advice on how to be deaf from the very people who knew most about it.

The majority of deaf children are born to hearing parents, but their outlook in life is strongly affected by all sorts of factors – where they live, whether there's any Deaf community around them, how their deafness is accepted by others. Within the Deaf community, the merits and disadvantages of cochlea implants have been keenly debated for the past couple of decades. Some see the implants as a 'cure' for deafness and a pass into a speech-based world, while others mourn the erosion of Deaf culture and the labelling of deafness as an 'undesirable' trait. Having won a long-fought battle for acceptance of a Deaf identity, some see the implants as defeat via the back door. Either way there's no question that at present there's a strong and healthy Deaf community in many parts of the country and that the internet has helped to connect many of those who would formerly have been isolated in more remote areas. It's also true that the NHS treats them as a distinct group with particular needs requiring specialist services.

But those who were born hearing and who consider speech to be their main form of communication are not part of any community. There are no pubs, no clubs and no separate services for them, and since the NHS deems the deafened capable of accessing mainstream healthcare, they're stuck down in the system along with everyone else. The deafened don't form strong bonds based on similar experiences and wander along the road for a pint together after work. Pubs hurt their ears, so instead they go home and see no one until Monday. Besides – as I pointed out to myself – what on earth would be the point of a deafened get-together? Most of those who lose their hearing don't know how to sign, so they have to talk, but the more people talk, the less they can hear. So what then? Does everyone just sit around in padded rooms twiddling their hearing aids and taking twenty minutes to figure a simple one-liner?

In practice, very few of the deafened ever learn either

British Sign Language (BSL) or any of its international vari-
ants. Like every other language in the world, BSL only works
when there's someone there to speak it back to you. So even
if someone who had been hearing and was now deafened was
to go to the trouble of learning BSL, the chances are that it
would leave them no better off because they'd most probably
still be living and working entirely within a hearing world,
and unless that person's family, colleagues and wider social
network all learned to sign as well, that person would still be
in much the same situation as someone speaking Ligurian in
Texas. Thus in effect the deafened become stateless; they're
no longer full participants in the hearing world, but they're
never likely to be part of the Deaf community either.

And when the inevitable happens and the deafened do
end up struggling, mainstream mental health services are
now so overstretched that those individuals would have to
be either psychotic or suicidal before they were referred for
treatment. Which means that in the vast majority of cases
something which may have started out as a relatively minor
issue – mild depression caused by hearing loss – often
remains unacknowledged and untreated. So the person's
confidence gets knocked and they find it harder to get on at
work, they lose their job or get shuffled down the organisa-
tional pack. They push aside their social life or forget about
having fun, and one day when there's no one left except the
dog and a great big unlifting lump of sadness, then at the
eleventh hour they finally come to the attention of mental
health services because they're standing in the middle of
Tesco's produce aisle yelling about elves.

AFTER A COUPLE OF YEARS of making do with the NHS's
analogue hearing aids it became clear that if I was to have
the best chance of hearing well, I was going to have to go
digital. As Steve had pointed out, digital aids weren't then
available on the NHS, so if I wanted them, I would need to

go private. Then and now, good privately fitted hearing aids are expensive – somewhere around £2,700 just for one, more like £6,000 for two plus the cost of the appointments and ongoing maintenance.

But in 2001 I plundered my savings and – after a further examination – was referred to an audiologist called Jacqui Sheldrake who has been working with the Deaf for so long that she now naturally enunciates with precision, giving each word its proper space. Once every few months I'd make the trip over to her and sit in the waiting room while the lines of sunlight crept across the carpet. Every year or so she would rerun the pure-tone tests and adjust the frequencies on the aids according to the degree of loss. Every year, I would watch the line on the audiograph move a little farther downwards, and every year I would watch the level of amplification keep pace in the opposite direction.

The aids Jacqui fitted me with were relatively discreet. If I was standing face-on to someone they probably couldn't see I was wearing them, especially if I had my hair down. And because this time around I was wearing them all day every day, I adjusted to them quickly. Having performed its clever neuroplastic magic trick, within three months my brain was already comprehending the filtered digital sound I was receiving as almost-true sound. Once in a while, something would remind me that what I actually had was something akin to an old cassette tape with the volume jolted up, but most of the time the aids just felt normal – so normal I rarely thought about them.

DR SALLY AUSTEN is the Consultant Clinical Psychologist at the National Deaf Mental Health Service Unit at Birmingham and worked at the Royal Throat, Nose and Ear Hospital in London's King's Cross for four years before moving north. At King's Cross, she was working as a clinical psychologist within an audiological unit treating both the deaf and the deafened, but in Birmingham she's working in a psychiatric unit in which the majority of patients have been deaf since birth. The distinction is critical. Whereas in London she was dealing with both groups, Birmingham is for the 60,000, not the 11 million.

The first time I meet Dr Austen is at her home near Bromsgrove. She is blonde, recently turned 50, and doesn't correspond in any way to whatever vague prejudices I've got about psychologists (dark, elderly, Hampstead). Instead, she lives in a space overpouring with children's drawings, boots, dog clobber, photographs, food and family life. It's a warm spring day and when we sit in the conservatory to talk, there's something obviously characteristic in the fact that she makes coffee for me but in an ice bucket by the windows there's a whole stack of other drinks – orange juice, Coke, cold water, milk – in case I should want those instead. She folds her legs up on the sofa and talks animatedly for an hour and a half without slipping into medical obfuscation. After a short while it becomes clear that she is both very good at her job and very humane in the performance of it.

It's taken me a very long time to get to this place. It's almost two decades since I started to go deaf, and now – long after the event, and long after my hearing has returned – I have come looking for answers. Now, I want to know what happens to the people who, like me, find that deafness has crept in and overturned their life. I want to understand.

When Austen was working at King's Cross, many of her patients were beginning to lose their hearing and were dealing – or not dealing – with the psychological consequences of that. As I discovered, there really aren't many people out there who have reacted to a diagnosis of serious hearing loss in a balanced and socially approved manner. Was there a common reaction amongst the people she saw? 'Yes. Depression. And social anxiety.' Both of which were exacerbated by isolation. 'If you've got a spinal injury – and I did work briefly in Stoke

Mandeville – you have a very powerful experience with a group of people who are all in the same hospital for a long time. So by the time you come out of there, you know what others are experiencing, you've got an identity, you've got a shared community. But with people losing their hearing, they go to the hospital, then they go away and attempt to hide. And the deafer they get, the more they'll hide.'

Austen has got used to watching for the ways in which their lives change. 'I can predict, for instance, what a deafened person will order in a restaurant.' Like what? 'It would probably be pizza. Things that they can be absolutely sure they know, they're not going to go for the complex thing. People losing their hearing lose the variety in their life. They'll go for the safe option – everything's safe, because otherwise it's a hassle. So say if they ordered steak, they know the waiter's then going to ask them how they like it done. It's got to be whatever is on the menu that's completely non-negotiable, non-discursive, because they just don't want to get into a conversation about it.'

When Austen was able to work with both deaf and deafened, there were things she could do to acknowledge that both groups had very particular issues and to work with them. For those who were losing their hearing, the most effective of all – and the cheapest – was just to listen. 'Us humans rely on communication, and communication is almost always verbal and audiological. If I could cure the world of loneliness, I would see ninety per cent fewer people.'

At King's Cross, Austen got used to being able to effect real change – not healing people exactly, but definitely allowing them to feel better about their situation. Now, because of pressure on resources and that distinction between the deaf and deafened, most of her patients have fallen out of the system, so what remains is only the most intractable issues. Faced with so many more urgent demands on its attention, the healthcare system regards this particular branch of their services as a backwater – it's the common cold of clinical psychology. Everyone wants to work with children or criminals, and nobody wants to spend their daily lives repeating 'How are you today?' at ten times normal volume.

I first came across Dr Austen not through the Birmingham unit, but because her name was on a list of contributors in a book on pseudohypacusis, or non-organic hearing loss (NOHL). In unclinical terms, pseudohypacusis is deafness with no apparent biological cause and ranges through a spectrum of issues from acutely traumatised individuals whose capacity to hear has temporarily shut down to those with 'conversion disorders' (those who have converted a psychological issue such as depression or stress into a physical symptom) to fakers on the hunt for compensation.

Probably the best known form of pseudohypacusis is amongst those who have heard something they find so horrifying that their brain has temporarily suspended its capacity to hear. During World War I large numbers of soldiers started coming forward claiming that they had been deafened. When tested, there appeared to be nothing wrong with their ears, and yet there was no question that they were unable to process or respond to acoustic stimuli. Many had also become mute. They weren't malingering, they were just so traumatised that their minds had slammed the gates shut and now refused all sound.

At the time the attitude amongst psychiatrists and army doctors in both Britain and Germany was that anyone who couldn't prove an injury or illness clearly didn't have one and must therefore be making it up. The issue, the doctors believed, was not the horror of war, it was a lack of moral fibre in those particular individuals. Besides, so many aspects of 'hysterical deafness' seemed counter-intuitive. A significant proportion of those coming forward were not the individuals who had been exposed to the worst aspects of battle. During World War II, many of those assessed by the US army as having psychogenic deafness had no combat service at all and were simply soldiers who had found the whole experience of war (the separation from their family, the army discipline) so difficult that they had lost the capacity both to receive and to transmit. Not that they received a particularly sympathetic hearing – at one stage, the US army's standard treatment for sufferers of psychogenic deafness was sodium pentothal, a chemical truth serum designed not to address the cause of the trauma but to force the sufferer into confessing that they had made a falsified claim.

I hadn't realised that there could be situations in which the human brain could just down tools and walk out on strike. I didn't know that, *in extremis*, all of us have the capacity to withdraw from one of our own senses. I thought that hearing was involuntary, like sneezing – if a sound was audible, then you had no choice about whether it entered you or not. But behind all those wartime accounts of 'hysterical deafness' there seemed something very truthful in the body's acknowledgement that sound – even sometimes just the sound of another human voice – can occasionally be terrifyingly powerful.

NOHL is not restricted to soldiers and can still be a risk for any victim of trauma. Modern psychiatric approaches would probably involve addressing that trauma and then establishing whether the deafness had improved as a result, or in examining the situation which caused the trauma in the first place. If it was connected to one single, shocking event, then temporary deafness may also have been a clever piece of biological prioritisation. If, for instance, you're involved in a car crash, then you may well remember the few seconds around the impact with extraordinary visual clarity, but also as an experience completely without sound. At that instant of greatest emergency, the brain has focused in on the senses it most needs, channelling everything away from hearing and into different parts of the body.

But pseudohypacusis is not only caused by trauma. The really surprising thing is just how much time it takes up – Dr Austen calculates that, at any moment, between one and three spaces in her Birmingham ward are occupied by patients who may not be as deaf as they say they are. Of course, someone who can hear fine when they're out on the street but seems suddenly to have developed a hearing issue when they get to the audiology reception desk is easy to spot. But other forms are not quite so easy, and however much audiologists are on the lookout for pseudohypacusis some do inevitably slip through the net. Because so much of hearing takes place in the brain and because there are many conditions like tinnitus which are totally real and often very traumatising but for which there is as yet no physical test, how exactly are you supposed to pick out one (genuine) person who doesn't respond to pure-tone tests or to normal auditory stimuli from another (malingering) person whose hearing tests come up just the same?

So is this pseudohypacusis thing really that common?

'It is fascinating, isn't it? I've had quite a few clients who have deafened themselves. Or have feigned their deafness, but I certainly have a couple who have literally deafened themselves.'

She was prompted to look into the issue in 2003 when she was approached by a worried colleague. 'An audiological scientist came up to me and said, "There's something I'm really worried about – I think we've just [cochlea-] implanted somebody who wasn't deaf. And I don't think they're the only person coming into our department who isn't deaf. Or not as deaf as they appear."' Over her next study leave, Austen read everything she could get her hands on about the subject – which wasn't that much, both because the issue was considered so niche, and because no one could ever agree on a definition of terms.

Working with another colleague, she came up with a diagnostic model, and then tried broadcasting that model to her peers. It proved a harder sell than she imagined. 'At the time that we wrote this, people were saying, "Yes, really interesting, but we haven't got any of those patients, thanks." But as people began to understand the concept and what to look out for, they were finding them all over the place, and it's now getting alarmingly big.' But why? Why would someone do that? After all, you've got millions of people who really overwhelmingly don't want to be deaf, and then you've got a few who just can't wait?

Austen responds by pulling out an A4 notebook and drawing a big square with two circles on the base line. The one in the left corner represents intent, and the other on the right represents lack of intent. On the vertical axis, she draws a line with one end representing two different types of gain – practical or psychological – and at the other end of the line, no gain. She points to the circle representing intent/ practical gain. Those are the malingerers who would prob- ably be after financial compensation and thus would only be deaf for the duration of their audiology appointment. Then she points to the other corner, the one representing lack of intent/gain. 'The other extreme are conditions like PTSD, conversion disorder where because someone can't deal with what's happening to them psychologically, their mind translates the trauma into a physical ailment because physical ailments are somehow easier to deal with. And then in the middle is this whopping great group which is facti- tious – that's where it is intentional but it's almost become habit. Like, "My life is rubbish if I don't do this, but my life is better if I do do this." And that might be attention- seeking – people like the attention, they like the company of the doctors, they're pathetically, horribly lonely so they keep going to doctors' appointments. And they like the process, the company, the attention. At the extreme would be Munchausen's Syndrome. So somebody who turns up for a cochlea implant – that's really quite Munchausen's.'

But what about the deaf wannabes? Out there, should you be interested, is a whole world of people whose major life goal is either to pass as deaf or to deafen themselves. Look online and there are plenty of sites in which people discuss passing as members of the Deaf community by learning sign or lurking around Deaf clubs and groups. There are others on which the best ways to induce hearing loss are discussed – perforating eardrums, using extreme noise to damage hearing. On both, the tone is either practical ('this is how you puncture an eardrum'), or lascivious ('turned on by the feel and sight of myself in hearing aids'). And even within the wannabes, Austen sees a lot of division.

'Once they became more visible on the internet it became clear that it wasn't just people who wanted attention within this group, there were all sorts of people. What's interesting about them is that they're a relatively small group, and because they're small they have to put up with each other despite very different motivations. So you've got people with a fetish thing going on who either find deaf people sexy or hearing aids sexy, you've got people who are quite autistic or on the autistic spectrum for whom noise and communication and sociability are just so miserable that they'd actually just rather be deaf or deafened or partially deaf because it gives them control over the communication space. You've got people who are psychopathic who somehow enjoy beating the system. They're the ones you can see on deaf forums going, "Ha! Did you know that if you do four hours of noise-induced hearing loss at 85dBs, then you get it past the audiologist!" Then you've also got people with a body dysmorphia, so in the same way as those who are transgender and who feel that they're born male in a female body, you can also get people who are born hearing but they feel they're a deaf person in a hearing body. So people's motivation is very different.'

How easy is it to spot pseudohypacusis?

'In terms of me personally identifying them it's really quite easy because the behaviours associated with deafness – the way someone holds their head, or the way someone answers and communicates, the tone of their voice – are predictable. Based on the tone of someone's voice, I'd be able broadly to say how long they'd been deaf, or at what age they were deafened. But my job is not to confront or humiliate. My job is just to think, "Well, something has made this person want to arrive here to tell me that, what is it that they actually need help with?"

'In the early days I used to try and cure people of their non-organic hearing loss, and I'd try and find all sorts of complicated graceful recoveries, and no one ever took me up on it. Whereas because the person might have multiple non-organic things as well – so a not-quite-real backache or a not-quite-real epilepsy or a not-quite-real something else – what I find is that if I work with the person towards accepting who they are, then all those other problems tend to drop off.'

5

Conduction

I SOON DISCOVERED that an admission of deafness provoked a variety of responses – sympathy and interest from many, and a sincere desire to make things easier, but also a saggy, dated humour. I made the quips easier, told them first against myself, got the hardest cut in fast.

Friends and family seemed confused. Face to face, nothing seemed to have changed. As long as the conditions were right, I still heard, so I still responded. Which meant it was a hard thing for others to judge – did that mean this was a big deal or a little one? What were they supposed to say? After all, this really wasn't supposed to be a problem for our demographic. We were media types in our twenties equipped to deal with specific types of crisis (minor wars, romantic calamities), not the stoppage of significant bits of ourselves. Once this had got past the headline moment, what exactly was the right emotional tone? Was this thing boring or dramatic, or could it safely be downgraded to yesterday's news?

It also took me a long time to realise that others were waiting to take their cues from me. On some secret level I might believe I had brought deafness upon myself, but I also thought I was the victim, which meant I also thought it was the world's job to be nice to me. In fact, as anyone with a health condition or disability knows, it's up to them to take charge. When I met someone new it would have been so easy just to say, 'I can't hear very well, so it would really help if you could speak clearly, keep facing me, and please don't cover your mouth.' By doing so I would have given them permission to ask, established the boundaries and made the situation much easier for both of us. Instead I made sure people couldn't see the aids and tried my best to pretend I could hear well. What that meant in practice was that I'd occasionally blank someone mid-sentence or walk away from them while they were still talking. Several people told me later that I'd ignored someone who had just asked me a question or failed to respond when someone called my name. Being knowingly rude to someone is one thing, but to do it involuntarily seemed terrible to me. So I'd blame myself, go down a little lower, become a little less capable of speaking at all.

It was no good me complaining that my friends never cut me any slack when I never cut myself any. I wanted people to notice, I didn't want them to notice. I wanted mercy but I didn't want pity. I wanted to be the victim, but I kept insisting on independence. I wanted others to be clear, but I smudged the problem. Nobody could quite get a handle on this. Least of all, it seemed, me.

THE ONLY WAY I COULD THINK of to find a foothold on this precarious ground was to look for role models – famous people who had been deafened in later life, and to find out how they had coped. The trouble was, that route only led to one destination. Think of someone famous and deafened, and who do you think of? Beethoven. Ludwig van Beethoven, composer, genius, immortal. As a role model for musical prodigies, he couldn't be bettered. As an example of how to live your life once you started to lose your hearing, he was ... mixed.

Besides, with Beethoven there was an obvious proviso.

Anyone who starts to lose their hearing can stake their claim before their own personal gods or monsters that it shouldn't be them, that it should never be them, that they of all people are special or exempt or engaged in lifesaving work of national importance. Anyone can protest that when the genes were handed out they'd been waiting in the wrong queue. Anyone can stand in the Court of Special Pleading and argue that the other chap should get it instead. But the individuals who surely make the most compelling defendants are musicians. What can it be like to make a song and then find yourself locked out of it? How would it feel to compose something and so humiliate yourself trying to conduct it that the orchestra drives you from the room? Who would you become if your own playing had fallen so far out of tempo that it was now a joke, an embarrassment, a source of public pity? What must it be to have a thousand symphonies resounding in your head but to be unable to hear a single note?

In the summer of 1802, Beethoven's doctor recommended a period of country life to ease his health and give him time away from the pressures of city life. Sent to a house in what was then the small rural village of Heiligenstadt and is now a suburb of Vienna, Beethoven found himself alone. He was 32 then and had been aware that he was losing his hearing for a while.

On 6 October 1802, he sat down to write to his brothers Carl and Johann:

> Oh, you men who think or say that I am malevolent, stubborn or misanthropic, how greatly do you wrong me ... I was ever inclined to accomplish great things. But think that for six years now I have been hopelessly afflicted, made worse by senseless physicians, from year to year deceived with hopes of improvement, finally compelled to face the prospect of a

lasting malady (whose cure will take years or perhaps be impossible).

Though born with a fiery, active temperament, even susceptible to the diversions of society, I was soon compelled to withdraw myself, to live life alone. If at times I tried to forget this, oh how harshly was I flung back by the doubly sad experience of my bad hearing. Yet it was impossible for me to say to people, 'Speak louder, shout, for I am deaf.' Ah, how could I possibly admit an infirmity in the one sense which ought to be more perfect in me than in others, a sense which I once possessed in the highest perfection ... If I approach people, a hot terror seizes me and I fear being exposed to the danger that my condition might be noticed ... what a humiliation for me when someone standing next to me heard a flute in the distance and I heard nothing, or someone heard a shepherd singing and again I heard nothing. Such incidents drove me almost to despair – a little more and I would have ended my life – it was only my art that held me back ... perhaps I shall get better, perhaps I shall not ...

The Heiligenstadt Testament has been read as lots of things – a suicide note, a will – but its emotional truth has never been doubted. Though Beethoven twists from shame to fury to grief to grandiosity and back again, it was not something written without forethought. Unusually, he made a clear and mostly error-free fair copy, meaning that he'd written at least one draft beforehand, and if it was ever intended as a suicide note he certainly never acted on it. The document was only found after his death in 1827.

Anton Schindler was the first and most unreliable of Ludwig van Beethoven's many biographers. His account of a rehearsal for *Fidelio* in 1822 should probably be taken with a pinch of salt, but gives some sense of the truth of Beethoven's situation by the time he reached his fifties:

After several days of indecision, he finally declared his readiness to conduct the work, a deplorable decision on many counts … In the very first number, the duet, … it was apparent that Beethoven could hear nothing of what was happening on the stage. He seemed to be fighting to hold back. The orchestra stayed with him but the singers pressed on and at the point where knocking is heard at the prison door, everything fell apart. Umlauf [the other conductor] told the musicians to stop without telling the Master the reason … The duet began again, and as before, the disunity was noticeable, and again at the knocking there was general confusion. Again the musicians were stopped. The impossibility of continuing under the direction of the creator of the work was obvious. But who was to tell him, and how? Neither the manager Duport nor Umlauf wanted to have to say, 'It cannot be done. Go away, you unhappy man!' Beethoven, now growing apprehensive, turned from one side to another, searching the faces to see what was interrupting the rehearsal. All were silent … I stepped to his side in the orchestra and he handed me his notebook … I wrote as fast as I could something like, 'Please don't go on. I'll explain at home.' He jumped down onto the floor and said only, 'Let's get out of here.' Without stopping, he hastened to his apartment … Once there, he threw himself on the sofa, covered his face with both hands, and remained so until we went to dinner.

Both doctors and musicians have spent the time since Beethoven's death squabbling over the reasons for his deafness and for his chronic ill-health. For every symptom there's been an explanation, and for every complaint there's been a cure. Over the past two centuries, Beethoven has been retrospectively diagnosed with typhoid, cirrhosis, syphilis, brain trauma, meningitis, lead poisoning, Paget's, Crohn's and Whipple's diseases, liver disease, TB, diarrhoea, diabetes, irritable bowel syndrome, anorexia, colic, jaundice, hepatitis, rheumatoid arthritis, labyrinthitis, gout, pancreatitis, lupus, sarcoidosis and smallpox – a whole encyclopaedia's-worth of sickness, disease and disorder abbreviated into one tumultuous life. He went deaf because he'd listened to too much music, or because he liked dunking his head in water to cool down, or because he'd chewed on his pencil too much, or because he got so angry he'd flung himself to the floor, or because he had once stood in front of a draughty open window with no clothes on. Even now, there are plenty of experts suggesting that if Beethoven had only stopped drinking or accepted the love of a good woman, then perhaps his story might have resolved itself very differently.

From his description of his symptoms, it sounds as if Beethoven suffered from an exquisitely unpleasant form of hearing loss – not just sensorineural loss, but also tinnitus and hyperacusis, which meant that the sounds he could hear were often painful to him. In the Heiligenstadt Testament he claims he had already been aware of problems for six years, though he also said he experienced periods when his hearing either stabilised or improved. Whether or not this was true his hearing gradually degenerated.

When it became evident that neither tepid baths in the Danube nor burning herbal poultices were going to restore him to health, he turned instead to mechanical forms of amplifying what remained. When that too failed, he began using the conversation books, quarto-sized notepads on which people could write questions or comments that he responded to verbally.

He also recognised what many musicians have understood since – that he could tune his own body. Though it's probable that Beethoven had also lost quite a lot of his conductive hearing, he could still rest his head against the piano to let sound play up through his skull, or use some form of conduit to draw it into himself. At the Beethoven Haus, there's a long wooden rod which the composer used to direct notes from the piano. Sitting at the keyboard with the lid open, he would clench the stick between his teeth and poke it towards the strings. When the wood touched the metal, the sound waves vibrated up through the wood to his teeth and into his bones. The technique wouldn't have given him a particularly good sense of the pitch of each note, but it would have allowed him some sense of its placing and its rhythm.

He was also supposed to have cut the legs off one of his pianos to better hear the resonance from it, though it's a hard story to verify. All his pianos would probably have had detacheable legs to make them easier to move, and though placing the soundboard directly onto a wooden floor would certainly have increased the 'feel' of each note, trying a run-through of the Eroica on a keyboard five inches off the ground would have been gymnastically near-impossible.

But when I looked through his story, what really interested me was what Beethoven made of his life once he became deaf. That, and what other people made of him.

ALMOST TWO CENTURIES LATER, I seemed to be playing a similar tune. Even if I couldn't follow the magnitude of his loss I could certainly recognise many of Beethoven's reactions, and those of others to him. How ridiculous it was to swear his brothers to secrecy in the Heiligenstadt Testament when it was perfectly obvious what was happening to him. How silly to talk to no one about it and then complain that the world didn't understand. How absurd to hurl his mechanical aids across the room because they could never offer him the sound of true music. And how desolate to stand beside his friend in the autumn fields but be unable to share a single moment of their experience.

And those flickering moods – I knew those too. *I'm such*

a victim, I'm such a martyr, I'm so lost, I'm so ashamed, I am wronged, the world's against me, maybe there's a cure, probably there is none. Reading the Testament back, it sounded like drink, though it wasn't. That was just what happened when you sat down, thought about deafness and allowed yourself a whole evening of feeling sorry for yourself.

Oddly enough, I did understand a tiny portion of his feelings about music too. I wasn't a musician, I was just an ordinary member of the audience. But the idea of a life without music seemed like the idea of a life without sky. I might not speak melody the way a composer did but I simply couldn't comprehend the notion that it might now be a finite resource. I tried, but I just didn't understand what that meant, and however long I thought about it I just couldn't make the concept compute. So soon there would be no more songs? How could music just cease to exist? Where did it go? And, more importantly, what else might then push in to fill that void?

I suppose it was an odd reaction. If I was going to lose my hearing, then surely music shouldn't be top of that loss. That place must surely belong to voices, to speech and laughter. But there was something really specific about that soft repeat-to-fade – something which didn't apply so much to voices, or which I could somehow handle better. In part it was because the difference between raw music and digital music was so stark. By now, it didn't matter whether I was listening to the massively amplified soundtrack of a film or to a friend with a guitar, all of it had to be filtered through the digital aids, which meant that all sound already had half its colour washed out. The shape of the song was still there all right – Beck was still Beck, Bowie still Bowie – but the emotional jolt had gone. There on the radio was the song's skeleton – its height and shape, its familiar attributes – but its guts weren't there any more. I just couldn't figure it out. Music used to be a sustenance, now it was a nuisance. It was there, but it was lost. Something of it was still in the room but it was a wraith of what it had once been. The music which I used to love was … well, it was OK, but it was really nothing to fuss about.

It also now presented a serious timetabling issue. Because my capacity to filter and prioritise sound had dropped, I had to make a choice about what I could hear and what I couldn't, which meant that any situation in which music was playing in the background presented a challenge. I could still listen to it, but if I did, it had to be the only thing in the room. It couldn't just be there noodling away behind the sofa, it had to be a conscious choice. And muzac – a shop's ambient twiddlings, playlists in a café – was now something which scraped at my concentration and fought the person I was trying to hear. Instead of being able to select what I heard, it was now every track sounding straight over the top of another. Which somehow meant that this new zombie music all medicated down to a kind of digital half-life seemed to be coming at me with far more purpose than the rich, living version which had existed before.

WHAT EMERGES FROM all the reported accounts of Beethoven's visitors was how very uncomfortable everyone found his deafness. In eighteenth- and nineteenth-century Vienna, hearing loss carried exactly the same connotations as it does now, and many people seemed perplexed as to how to handle him. Should they simply deal with Beethoven the man and pretend the deafness wasn't there? Should they pity him? Or should they behave as if he was some dangerous escapee from a zoo, best tranquillised with flattery but contained behind musical bars?

The trouble was that Beethoven was famous – excessively famous – during his own lifetime. First he was famous because he was a brilliant composer, and then he was famous because he was a brilliant deaf composer. But even before he started to lose his hearing, he had become the subject of two powerful but conflicting myths. In one, he was a genius, an immortal adornment to Europe's creative pantheon. In the other, he was a nutter and a slob who could barely dress himself. It was Mozart all over again, a musician who wrote as if he was taking dictation from angels, but a man so backward he still thought fart jokes were funny. In Beethoven's case, the addition of deafness only strengthened both perceptions. He was a colossus, but he was unmanageable. He was a prodigy, but he was an idiot. He was a masterpiece, but you couldn't go near him. His music was celestial, but he raved like Bedlam. Very few people could line up the two images – half-genius, half-miscreation – and steer a clear course between them.

Beethoven was used to – and often complained about – a constant stream of musicians, acquaintances, friends, tradesmen, conductors, petitioners, patrons and students appearing at his door. In the years when he was still able to hear shouted conversation, some visitors wrote up their meeting as if nothing was amiss. OK, so the composer might have had his shirt on upside down or might not have been wearing any trousers, but they'd talked about music and politics and Great Things, and the visitors had returned home delighted to have experienced a real-live prodigy in his native environment. Or they'd found him in a state of complete disorder, as Friedrich Starke had done in 1821:

> I called on him one morning and being a friend was at once admitted by the servant. I searched through the house in vain until I reached his bedroom. As knocking was useless, one generally entered unannounced, and to my astonishment I found Beethoven sitting in the middle of the room in his shirt ... He had thickly lathered his face the evening before, and had forgotten to shave. His absence of mind was a ludicrous and yet at the same time a melancholy spectacle. The soap

had dried overnight, and looked like paste sticking to chin and cheeks. But not at all disconcerted, he only motioned to me to withdraw until he had put himself to rights and came out of the room in a few minutes.

There are certainly countless stories of Beethoven's indifference to dress or behaviour. On a couple of occasions he was mistaken for a tramp, on another he was taken into police custody for causing a public nuisance. Since he was always having tantrums and sacking his servants, his clothes often went unwashed for weeks at a time and his friends finally realised that the only way to get him to freshen up was to steal the old linen, and substitute new. Beethoven apparently never noticed the difference.

All of which ties in perfectly with the great romantic notion of genius – of individuals whose mind and spirit are so utterly directed towards their art that there remains no space for anything else. Thus the general view of him was summed up by Carl Friedrich, Baron Kübeck von Kübau: 'Whoever sees Beethoven for the first time and knows nothing about him would surely take him for a malicious, ill-natured and quarrelsome drunk who has no feeling for music.'

For his musical contemporaries, Beethoven's high profile presented additional problems. With time and teamwork, they were able to come up with strategies to circumvent some aspects of his deafness, thus allowing him to continue appearing in front of audiences. Though Beethoven himself understood that his hearing loss prevented him from conducting his own works, the desire to present them never left him. On the one hand, his contemporaries understood that desire – even encouraged it, since the presence of the composer at one of his own performances was a huge publicity draw. On the other, and as Schiller's account of the *Fidelio* rehearsal had revealed, if Beethoven conducted, then the orchestra couldn't play.

Part of what is interesting about the responses of Beethoven's visitors is what they say about attitudes to deafness. Admittedly, Beethoven was an extreme case. Much of the time he probably was unmanageable – a rage of different passions and conflicting drives – but some of that anger must have stemmed from the frustration in communicating to the outside world who he was. Even so, what his visitors' accounts often highlight is just how little has changed in the centuries since. Evident in every single line of all those accounts are the personalities of those who met him. His visitors didn't just observe him, they brought to the room their own confusions, prejudices, jealousies and desires. And what they saw usually says far more about them than it does about him.

6

Rock

IF I WAS HAVING TROUBLE listening to music, then what on earth must it be like to know that it's music itself which has made you deaf in the first place?

For both classical and rock musicians, there is an entirely genuine threat posed by getting too close for too long to the thing they most love. In the last four-hundred-odd years since human beings started gathering songs into groups and ensembles, the risk of deafness has remained their shadowy accompaniment. One recent German study put the risk of hearing loss as four times greater among professional musicians than it is for the general public. Amongst the classicists, many violinists damage their hearing in the left ear – the side closest to the body of the instrument. French horn players can not only deafen themselves but those positioned closest to them. In fact, it's orchestral musicians generally who are most at risk of hearing loss since they may well be placed physically closer to another instrument and expected to rehearse and perform for far longer than most rock musicians every day.

In the rock world, the problem is not so much sitting for long periods next to loud noises as being surrounded by very high volumes. The majority of club sound systems operate at over 90 decibels – somewhere between a power drill and a chainsaw. Return to that early image of the mechanism of the ear – the eardrum, the middle ear, the cochlea, the auditory nerve, all of them operating with a lapidarist's precision – and then imagine wave after wave of overpowered sound banging like a lump-hammer onto the tympanum. It's not the frequencies that are doing the damage, it's the motion. Hit anything hard enough for long enough, and of course it's going to break.

Everyone from the World Health Organisation to Zane Lowe has started issuing warnings about the potential epidemic of hearing loss among music lovers. In 2015 the WHO produced a report suggesting that 1.1 billion people are at risk of permanently damaging their hearing by listening for long periods to over-amplified music. Half of 12- to 35-year-olds in richer countries are exposed to unsafe sound levels through personal audio devices and 40 per cent are going to gigs or festivals where the sound level will almost always exceed safe levels. The WHO recommends that we should all reduce the volume, give our hearing a break by stepping out of noisy environments every few minutes or so, and wear earplugs – better still if they're the ones designed specifically for music and which have some filtration capacity.

But there are flaws in this plan. Firstly, teenagers may be unaware that loud music can be permanently damaging and are often resistant to wearing earplugs. Back in 2005/06, AoHL surveyed a group of young clubbers and found that 70 per cent of them had experienced ringing in the ears after a night out, though almost none of them realised that this was a warning of potential future problems. Besides, if you're at a gig and you've just spent three hours worming your way to the front of the stage in order to catch every fleck of spittle from your favourite band, then nothing short of forceps is going to get you out again.

And secondly, musicians play it loud because we like it loud. There's a lot of music out there which just isn't designed to be quiet. Punk, dub, drum & bass, hip-hop, rap, house, techno, metal, old-fashioned rock'n'roll – all of them were born to be heard at high volume. All were supposed to kick against the rhythm of a heartbeat or thrill you up through the soles of your feet, and many music lovers would probably say that if you take the noise out of music it stops doing its job. The Sex Pistols did not write 'Anarchy in the UK' to be heard at the WHO's recommended safe noise level, nor the White Stripes howl America down for an audience of HSE-approved earplugs. A live experience has the ability to lift people because it's right there banging against their bones, and even music which sounds as if it's supposed to be heard quietly – the work of Palestrina or Arvo Pärt, say – is still designed to be heard just as loud as the unamplified volume of several dozen human voices.

IT TOOK ME A LONG, long time to find anyone from the rock world who would talk to me about hearing and music. When I started this book I searched for musicians with hearing problems and came up with what I thought was a fairly extensive list – not just older suspects like Pete Townshend or Gary Numan but those in their twenties and thirties, many of whom may have developed tinnitus. In fact, there were so many people prepared to declare themselves sufferers it often seemed as if there were a far larger number of musicians out there with deafness than without. Great, I thought, this shouldn't take long.

But trying to get them to talk on record was a different story. Over the next few months I approached everyone on that list from Jazzie B to Plan B. Nothing doing. They might all have been enthusiastic supporters of charitable campaigns in theory, but in practice they were all amazingly busy on new albums. In the end I tried over thirty musicians in both Britain and the States, but I've had more luck persuading unconvicted criminals to talk to me than I did rockers with hearing loss. I began to get an inkling of what 1950s Hollywood or 1970s Westminster must have been like – an entire professional class still locked in the closet.

OK, I thought, if musicians themselves won't talk, then perhaps the people who make those musicians' music work will. In March 2016, a friend texted to say he'd just heard a radio obituary of George Martin during which they'd played a clip of him talking movingly about his deafness. His son Giles is also a producer, so I got in touch with his management. Sure, they said. Would half an hour at the studio be OK?

Take the word 'Beatles' out of Giles Martin's CV and he's a successful record producer who has worked with artists including Jeff Beck, Elvis Costello, Kate Bush and the fastest-selling classical musician of all time, Hayley Westenra. He helped put together the music for the Queen's golden jubilee and the London 2012 Olympics, he's an instrumentalist and a producer of audio-visual spectaculars, he's composed film scores and assembled soundtracks. But put back that word and you get a whole different story. Search for Giles Martin online and he becomes less a person in his own right and more a piece in a much bigger machine. Or just a familial connection; the son of the Fifth Beatle. Or – as the *Telegraph* would have it – the music-producer son of the greatest music producer who ever lived.

The studios on Abbey Road are in the process of being refurbished so when Giles Martin appears he leads me over to the building next door, which is currently providing a temporary home for his own offices. Somewhere outside the window a perpetually self-renewing stream of tourists and trippers freeze mid-step on the famous zebra crossing, a thousand families recreating the cover of the *Abbey Road* album a thousand times a day.

Giles is tall and well kept with thick swept-back hair and a particular stillness to him, a kind of caution, as if some part of him is always listening for trouble. His hearing, incidentally, is fine – more than fine, in fact; it's in demand, and very expensive. In addition to his production work, he's an audio consultant to Sonos, an audio home-entertainment company, and is occasionally called in by other specialists to provide an expert overview.

His father's death was still recent and as he talked about him he flipped between present and past tenses. His bond to George is written all over him, but so is the conflict between honouring that bond and being his own man.

Giles is 47 now, though he was still a child when his father first started to lose his hearing. As George Martin explained in later interviews, the first he knew of it was some time in the late 1970s when he was in his own studios at home. In the control room the sound engineer was running a series of tone tests in preparation for a recording session. As he spooled through the tapes, George found himself watching the dials. He could see that all the needles on the dials were twitching to the right, which meant that sound should have been filling the room. He could hear nothing. '"Bill," I said, "what's the frequency you've got there?" He said, "It's twelve kilohertz." I said, "Oh shit." I knew I could no longer hear 12 kHz and never would again.'

To start with, George Martin lost the treble, though. 'He ended up losing all his hearing almost up to a kilohertz,' says Giles now. Severe deafness would be between 70 to 90 dBs, or around 1 kHz. 'That much. All frequencies. Like a filter. He became *very* deaf.'

Several decades after that first hearing test, father and son tried again. 'We used one of the studios downstairs and I was being deafened by things that he couldn't hear. Didn't even have any idea that they were on.' So it wasn't like he was hearing some frequencies but not others? 'No. They just didn't exist. So beyond a certain point hearing aids didn't help because they're just amplifying sound, and as his hearing became worse there were no receptors to pick up those frequencies.'

It took a long time before George Martin went public. Though the Beatles had already split when his hearing started to decline, he still packed in a further three decades of work before finally retiring in 2001. In addition to the 30 number ones he'd had with John, Paul, George and Ringo, he then produced or contributed to everything from Bond soundtracks to Elton John's multimillion-selling tribute to the Princess of Wales and co-wrote a memoir rather poignantly entitled *All You Need Is Ears*. Throughout all that time, his hearing continued to worsen. It was a loss which he knew had been caused by years of sitting in studios listening to music at high volumes. 'Self-abuse,' he called it later, castigating himself for not realising the consequences of prolonged exposure. 'Is your hearing loss a result of listening too carefully to loud music?' he was asked in a 2012 *Jazzwax* interview. 'No,' replied Martin. 'It's because I wasn't listening carefully enough.'

'Not until very late on in life did he talk about it,' says Giles now. 'He died at the age of ninety and you won't find an interview with him in his seventies talking about it.'

Initially George had not wanted his son to follow him into the music business, though by the time Giles was sixteen or seventeen his father had started asking him along to meetings or sessions, ostensibly as work experience but really in order that Giles could translate. 'He didn't tell anyone he was going deaf because he still wanted to work. And who wants a deaf record producer? So I was my father's ears. And we became very close because if you're someone's ears, you learn a lot about a person even if they are your father. And I'd be this person hanging around with him and people would think, "There's that precocious son, why is he bringing his son along?"'

So what did that mean in practice? 'To begin with, when his hearing was failing – so this would have been when he could still work – I would have been about seventeen or eighteen, he would say things like, "I've got a feeling that this record doesn't sound right." So, for instance, the red and the blue albums, the famous Beatles' greatest hits albums from different eras, they came out on CD, and he sensed that they sounded too bright. That would probably have been 1990 or 1989. And I came in here to check, and I remember the guy absolutely hating me for being here because I was this eighteen-year-old coming in and saying, "It doesn't sound right," because it didn't. I compared the CD to the vinyl, played the vinyl, played the CD, played them again, and I said, "They don't sound the same, and they need to sound the same." So you learn what the other person wants to hear. And it's not as though my father was gelded completely by his hearing. He could still do lots of things. He was still a very good musician and he could still write and arrange and all these sorts of things, but yeah, he would present me as being a trainee, which I was, but ...'

So for a long time, you were keeping his deafness a secret between the two of you? 'Yes. And then as people got to know us, they would know what was going on or ... just accept it.'

Were you ever tempted to misreport anything?

'No! No. What I wanted to do was make sure my dad didn't come across as looking like an idiot. Because he wasn't an idiot, he was just losing his hearing.'

But to be that guide – that's a difficult position to be in.

'Yes, but you learn a lot.'

Were there moments when it was too close for comfort? It's an almost intimate role.

'But it's your parent. You can be intimate with your parent. And he couldn't have had anyone else do it, that's the thing. It was funny – I lived my life a bit like Benjamin Button with my dad where, by the time he died, we were incredibly close. I was probably one of the closest people to him, if not *the* closest person to him. Because we're very similar, and we have the same talent, and also I was his ears.

It was a weird situation. I now have a rejected guide dog for the blind as a pet, but I was a hearing-aid person from the age of fifteen.'

On a practical level his father was able to make some changes to improve things, but emotionally it was a different matter. 'There was nothing for him. There was no bonus or added experience about becoming deaf whatsoever. It was just darkness. And that sounds terrible to say, but to him … It was this thing that, above anything else in the world, was the one thing that he would have changed. Absolutely. Totally. Because it was … his powers were taken away from him.'

Is that how he felt? That his hearing wasn't just a sense, it was a power?

'Yes. I think so, yes. Because it was an ability. One of his abilities was that he had very good ears and suddenly he couldn't hear. He found it very frustrating. He could sense things but he couldn't hear them.' It was that familiar thing – some frequencies declined quicker than others, which meant that in the studio his ability to listen to a track and spot the defects or omissions also declined. Having once possessed perfect pitch, George was now groping around for mislaid notes.

One hearing specialist who sees a lot of musicians says that while many of the issues they grapple with are common to anyone with hearing loss, there are also some which are particular to music. Often, musicians are dealing with several interlocking status anxieties. Many are too famous and well established to worry about being cool, so that's not a problem. A few of them worry about being seen as vulnerable and dependent, so that's sometimes a problem. But the overriding concern for almost all of them is the association of deafness with age, there being nothing quite so noxious to rock as oldness. Which means that on the one hand they're proud they've lost their hearing because it testifies to a life spent smashing sound barriers, but on the other they're worried that admitting it means that they've finally lost it.

For record producers or engineers, however, it's a different story because the whole of their professional reputation is built on the foundations of one single sense. Giles likens his musicality to a palate. 'So much of what we do is to do with balance and taste. If you imagine making wine, it's complex and has acidity and strength and fruitiness or whatever they call it, but there's a balance to getting it right. And music is to do with taste, but also balance – you need to balance the right frequencies. If you balance the wrong frequency you'll have a record that sounds too tizzy or too muffled and boomy. And my father always made good-sounding records.'

But the thing was, it wasn't music that his father missed most. 'People would say to me, "It's so awful, your father's hearing loss. He must miss music so much." And you'd go, "He doesn't give a *toss* about music." He didn't. He'd heard enough music to last a hundred people's lifetimes. It was all the other stuff. The sad thing for him isn't that he lost his hearing when he had such a huge talent for music, the sad thing for him was losing his hearing because he wanted to talk to people. The rest of it is just showbusiness bollocks, it really is.'

What do you think are the emotional consequences of deafness?

Giles responds without hesitation. 'Fear of isolation. That, above all others. I think isolation is the key word for me as far as hearing loss goes. My father always said that given the choice he would rather lose his hearing than go blind, which in some ways surprised me and in other ways didn't. He died looking at the tree in his garden from his bed, and loved looking at his grandchildren. I get it now.'

He tells a story from a couple of years ago when his father was seriously ill but still sometimes able to participate in family events. Giles had cooked Sunday lunch but could see that George was struggling, so he suggested he go upstairs to rest. His daughter Alice, who was seven at the time, tucked George in and pulled the bedclothes up around him. 'She kissed him on the forehead and she said, "Now maybe your ears will get better." Because for her that was the biggest thing. They knew about the sadness of his isolation, not being able to hear his grandchildren.'

Connection?

'Yeah. It's that.'

And humour.

'Yes. And being witty. And the fact that people think you're stupid. And he was very bright, and both he and I rely on wit and being almost acerbic at times because that's our nature, you know. We take the piss out of people, we take the piss out of each other, and we expect the same thing. And he would still be very funny, but he had to consider it instead of it being instant.'

Giles's voice is deep, which meant that George could hear him right up until the end. But with other family members, it was harder.

Could he hear your children?

'No.'

Your wife?

'Not as well.'

So women's voices generally …?

'No. Sexist.' He grins.

What about instruments?

'He could hear a drum because he could hear the low frequency of the beat, but a violin would be useless.'

How did other musicians respond to his deafness?

Giles considers. 'I think probably with fear,' he says slowly. 'It's like how we respond as humans. We've got to that age now, you and I, where friends get cancer and things happen, and our response when we hear about it really is fear. So if you meet someone who's a legend in music and who has made more records than most other people, and you suddenly find that there's a frailty to it all, I think the reaction is probably, "I hope that doesn't happen to me."'

And also a suspicion with many of them that it is happening.

'Yes. Without question. But it's how you deal with that. I saw someone in New York, a very very famous recording artist who I haven't seen for a long time, and I was backstage at an aftershow party. And as soon as I talked to him, I went, "Oh my God, you're deaf." And I said to him ... I said, "Are you OK?" And he went, "*Ooof!*" He lets out a long, troubled breath. I said, "This is awful for you, isn't it?" And he went, "Yeah," and I went, "I know, because I've grown up with this, and because you knew about my dad, you've worked with him, so you know I grew up with this." And he goes, "Yeah, I know." And I said, "It's just so nice to see you and I'm sorry you have to deal with this crap." And you're in this weird kind of isolation booth, and it gets tiring for the people around you, and you don't want to be that burden and all that sort of stuff. And you just wish that with a knife you could cut the bubble wrap from around you enough to hear anything – it's almost like you've lost a language.'

So, I ask, are you good at picking up the signs of deafness?

'Oh yes. It's like a gaydar.' He laughs. 'Absolutely, of course. I can know by looking at someone.'

How?

'By seeing how much someone is watching someone's lips move. I've grown up around it, and it's like anything else – it's not a huge talent, but I can spot it.' Which reminds him of a friend of the family who was deafened through meningitis at a young age and who now lives in New York. 'He's now forty-four years old, became a kite-surfing champion, and women think he's the most unbelievable man because a) he's incredibly good looking, and b) he's incredibly engaging, because people don't realise he's deaf. He looks so intently at people. And I remember saying to him, "Geordie, it's extraordinary – these women just love you. They think you're so deep, but you're just shallow and deaf."' He laughs again and adds, 'It's true. They say, "My God, he's really interested in me," and you go, "No! He's just struggling to hear what you're saying!"'

So, as well as getting very good at spotting the physical signs, it seems like working with George also made him an astute psychologist?

He considers. 'Yes. But there's so much more communication than being able to hear. We sense so much. It's like Stevie Wonder – I went to interview him with my dad in my early twenties for this programme called *Rhythm of Life*, and he was extraordinary. He sat at the piano and he almost scored his conversation. He constantly played the piano, played along with his conversation to embellish what he was saying. It was the most bizarre connection that he had with music and speech.'

So presumably he knows a lot of deafened musicians?

He looks down. 'I'm not in a position to name and shame people, but you would find that most touring musicians who are beyond the age of fifty will either have some or severe hearing loss.'

They won't talk about it to anyone? Or do they talk about it between themselves?

'Yes, they do, because it's apparent. It's very apparent when you talk to them that they're deaf.'

One estimate suggests that around 60 per cent of rock'n'roll's Hall of Fame have some degree of hearing loss or tinnitus, but it remains unlikely that any of those individuals will discuss it publicly. Just as in 1950s Hollywood, their reasons for not coming out on the issue are persuasive. Deafness is a difficult thing to admit to yourself, let alone an audience of millions, and besides, a decline in hearing doesn't mean a decline in ability. Musicians who have been playing for thirty years don't suddenly lose that capacity overnight; what they risk losing is their audience's faith in that capacity. So why should those musicians go public when doing so risks rendering them a professional write-off?

'I don't think that being deaf would make you a worse musician in any way,' Giles says. 'It might make you a worse recording engineer or record producer or a worse person at making speakers sound good, definitely. But it wouldn't make you a worse guitar player or drummer or bass player, or even singer.'

So what would deafness do?

'It would make you enjoy things less. The people I know who are suffering from hearing loss, the people that I've worked and recorded with, you wouldn't know that they were deaf by their performances. But you would know they were deaf by their conversation.'

Which presumably means that many musicians – Pete Townshend of The Who being one – go on playing live long after they became deafened? Townshend is one of the rare stars to have spoken openly about his hearing loss. 'I have terrible hearing trouble,' he said in a 2006 interview, and then, wrily, 'I have unwittingly helped to invent and refine a type of music that makes its principal proponents deaf.'

'Yes,' says Giles now, 'but Pete's not even that deaf. He's just talked about it, so he's known for it.'

Whereas you know people who are far worse?

'Yes. Absolutely, yeah. Like my father, for instance.'

As he explains, he's just been working on a single to go with the new Ron Howard film of the Beatles at Hollywood Bowl in 1965. A year later the band gave up playing live as the screams of their fans drowned out the sound of their playing. 'So at buildings like the Hollywood Bowl, they couldn't hear themselves. And I said to journalists, "If you have earplugs in and you try and play an instrument or sing, it's difficult. Singing's OK, but listening to your instrument is very hard. Doing it with two people with earplugs in is ridiculous, and doing it with four people with earplugs in – that's effectively what the Beatles did at the Hollywood Bowl. And yet they all sang and played pretty much in tune and in time.' In other words, the Beatles must have spent much of their live career playing deaf. They weren't actually deaf, they were just forced to behave as if they were.

But those who are actually deaf aren't going to stop playing either. So, I ask, you could have an entire band playing live every night, and each member would have some measure of hearing loss or tinnitus?

'I wouldn't say *if* you could do, I would say you probably *do* do. If you gave the Rolling Stones a hearing test …'

But they're in the age bracket where …

'They're in their seventies. People lose their hearing anyway. Fact. We're designed not for electronic music, not for amplification, not for headphones blasting out your ears really loudly, our ears are designed to hear a crackle in a forest.'

But why would some musicians be playing better with hearing loss than without?

'Because they've been playing for longer.'

So it's just muscle memory?

'Yes.'

OK, I say, still trying to puzzle something out, but why do so many of us have such a need for loudness? I know about protecting my hearing, I really do, but I still want to feel completely surrounded by the sensation of music even if I know that too much of that sensation is harmful.

'Because,' says Giles, 'you need that volume to create that excitement, and because I think it's something we don't experience. We like things we don't experience. We like going on a roller-coaster. We like to go fast, and then we want to go faster. We want to listen to music loud, and then we want to listen to it louder. We like one glass of wine and then we want four. We're drawn to excess, I think, and sound is part of that. If I played you two versions of the same song through exactly the same speakers and I audio-balanced them for you and one was a decibel louder, you wouldn't be able to perceive that decibel change, but you'd think one was better than the other, and it would always be the louder one. It's an interesting test to do. It sounds brighter, it sounds better – it's funny, but it's easy to trick ourselves. And it's the same thing with loud music as it is with smoking. We know it's pretty bad for us, but we still do it. When you turn up music, you want to hear that thud, that shake. That's what you get excited by, by being shaken slightly, and that's why you want to stand by the speakers. And that's why live music, music in the air, even if it's a symphony concert – it's still sound waves hitting you, and there's something about being there, and being bombarded by sound.

'The other thing about music and sound is this sort of wavering purple aura, this magical fragile thing.' Music being, of course, half science, all art, and pure enchantment. So while his own skill at listening is called on by audio boffins, what they want from him is not formulae or equations but his wizardish tricks with a hook and a bridge. And, almost certainly, something else. They want a little sprinkle of Beatles fairy dust – names, a few anecdotes, maybe just to stand next to someone who had stood next to Paul McCartney.

But from Giles's point of view, that proximity means he's often left dealing with a combination of envy and wish-fulfilment from others. Projects which had in fact either been collaborations between father and son or which Giles had worked on alone were credited entirely to George.

'People would come up to me and they would completely ignore age or disability because they didn't want to believe that was the case. They'd say to me, "What's your dad working on now?" I don't think I ever really got angry about it, but I redid the *Love* show in Vegas and Steve Tyler from Aerosmith was there. He came up to me and he said, "You're Giles," and I go, "Yes," and he goes, "Did you have anything to do with this?" I said, "Well, I made it." He goes, "I thought George made it." And you want to say, "Of *course* I did it. He was *eighty*. And he was deaf." But instead you go, "Well, we worked on it together, we worked very closely on this thing.'"

Was that lack of understanding about your father frustrating?

'No. I loved him.' His voice has risen.

But loving someone and finding it frustrating are not mutually incompatible.

'I think you balance things out. You take the rough with the smooth. It was just the way it was. I'm not sure I would have changed much.'

We're looping back to the beginning. Almost the first thing Giles had said was, 'People think people only employ you because you're the son of George Martin. I don't mean that in a sad way, but it's an inbuilt battle, your mechanism for not getting too cocky or too ahead of yourself.'

Is that a risk? That you might get too cocky?

'I think you can always get too cocky. I don't know. But the point is that I always approach any job or any situation by thinking, "Can I do this, am I the right person for this?" as opposed to, "Thank God you've come to me!"

'I actually said this to my father when he was very sick – I said, "Dad, did you ever think that you weren't good enough?" And he went, "Why? That's a strange question." I go, "Well, good enough at music." And he goes, "That's a *very* strange question." I said, "Well, because I often have that." He goes, "That's a ridiculous attitude. Why would you think that? I think you're brilliant at music." I said, "Well, thanks, Dad! But do you sometimes get asked to do something and you think, 'Oh my God, how am I going to do this?' And then you end up doing it and people say it's OK or they like it and you think, 'Well, I got away with that, what's next?'" And he goes, "No. I always thought I was brilliant!" He laughs, remembering. 'It was a very funny interchange.'

He checks his phone. He needs to go. If you can think of any musicians with hearing loss who might talk on the record, I say, then let me know. I've said it so often to so many people in the past year the words are just mechanical, and his mind is elsewhere. 'I'll have a think,' he says as we walk back down Abbey Road. 'But I already know what the answer will be.'

7

Acoustics

AS MY HEARING SLIPPED AWAY, I realised something odd. Until it actually started to happen I'd just assumed that, as I became deafer, things would gradually just get quieter and quieter, as if every year someone was sliding another layer of glazing into the space between me and the world. But this wasn't like that. Yes, the volume was definitely going down, but the odd thing was that deafness wasn't making me less aware of sound, it was making me more.

Having been indifferent to acoustics, I was ... well, not obsessed with it exactly, but it certainly seemed to occupy a lot more space in my life. Up until then I'd never really thought about the difference between the sound in a room with high ceilings and in a room with low ceilings, or why it was that some streets sounded softer than others. Did it really matter if you all sat in the big main room of a pub or squeezed yourselves into one of the snugs? In hearing terms, was there really much of a discrepancy between a double-glazed 1960s tower block or a suburban house near a busy road? Why should one large impersonal meeting room feel private while another sounded like drumming on dustbin lids? Was there any real difference between a floor of stone or of wood?

Previously I suppose I might have noticed that there was a difference between the way sound behaved in a cold winter hall or in a warm front room. But when I walked into someone's house to meet them for the first time, acoustics wasn't generally the first thing on my mind. Pre-deafness, I'd be thinking about the person themselves or what they'd said their son's name was, or – if I was interviewing them – what questions I needed to ask. Now, first thing, I'd walk into the room and start figuring out the ratio of tiling to lino or try to work out how I was going to get round the noise of their washing machine's spin cycle. I wasn't concentrating on them any more, I was concentrating on their environment.

I'd also started noticing that sound had different shapes. The way someone's voice sounded when they walked beside me down a street versus the way they sounded when they walked across moorland. The difference between a shopping centre's open spaces and its overstuffed shops. The funnelled passageways of the Northern Line or the vaults of Baker Street. DLR versus Overground. Victoria Line versus Central. Grass versus concrete, metal against glass. Which was easier: a conversation with someone near a school at breaktime, or the same conversation by a building site? Taxis, sheds, warehouses, stations, barns, old houses or new ones. Every single situation had its own particular sound, and that sound would alter depending on the number of people and the atmospheric conditions.

Exchanges with friends or colleagues started to take on a pattern. We'd agree on a date and a time and then everything would grind to a halt for a stand-off over destination.

'What about Mash?' they'd say.

I'd remember the scrape of crockery and the roar like an aircraft hangar. 'Sort of,' I'd say, 'or the local Chinese?', my thinking being that, OK, so the local Chinese might have been twice condemned by the Food Standards Agency, but at least it had carpet.

There would be a pause. 'The Eagle?'

Me: 'Or a deserted curry house in Hornsey?'

I'd sit there, fingers hovering over a two-sentence email,

wondering just how far I could push this before gaining a reputation for being a complete diva. How could I say I'd rather meet in a place with coleslaw and no atmosphere than somewhere where it might take me twenty minutes to figure out what the waiter had said? Was it better to go to a meeting in a noisy coffee chain and have people think you were a weirdo for staring fixedly at their lips the whole time, or confess and meet on a park bench near Bayswater like characters in a Le Carré novel?

It seemed like such a small thing but our priorities seemed to be pulling apart. My friends wanted to catch up and have something decent to eat. I just wanted to know whether they were speaking English or not.

IF YOU HAD HAPPENED to be in Glasgow in the spring of 2015, and if you had happened to be standing on the north shore of the Clyde near Glasgow Harbour, you would have been able to look over at the opposite bank and see a great steel shape filling the far right entrance of the big shed at the Fairfield shipyard. Except for a slight in-dipping as it nears the ground, the shape is almost square. Its sides are painted red and the face nearest the riverside is sliced by horizontal lines, each one representing another deck level. All over it there are people in overalls moving up and down the scaffold stairways trailing cables or angle grinders. The big shape slots into the shed's entrance so precisely that it takes a couple of moments before you realise it isn't just part of the permanent structure, an illusion reinforced by the scaffold stairs jumbling their way up the front and the shrink-wrapped spray booths barnacled to the shape's side.

The shape is a cross-section of an aircraft carrier, a slice of ship split right down the middle. It's just one section of HMS *Prince of Wales*, the second aircraft carrier being built at Govan for the Royal Navy. When it's finished it will function as a medium-sized floating airbase complete with hangars and workshops and a 280-metre flight deck tailing off over the sea. The first carrier, HMS *Queen Elizabeth*, is already over at Rosyth on the Firth of Forth being fitted out.

Over on the south side of the river the view is different. Once through the security procedures and the official preliminaries I'm given a set of white overalls, a pair of boots and a hard hat. My escort, Derek McCaffrey, the Operations Manager for the build, takes me out and round past the docks.

There's been a shipyard on this site since 1834 and during that time a whole township of roads and junctions has built up. Over there towards the roadside are the Edwardian management offices, now boarded up, and below our feet lies the lacework of cobbles and rails from the making of ships in the golden age. Back in the 1960s when Sean Connery was still licensed to kill, he made a documentary called *The Bowler and the Bunnet* on the bad blood between management and workers in this yard. One of the scenes shows him riding round one of the sheds on a bicycle, a big man made insignificant by the huge steel hollows. Painted walkways on the roads show people where to walk and where not to walk, and when work starts in the morning and ends in the evening, there's still the same mass movement of population heading towards the gates as there was back in the days when they built the SS *Stirling Castle*.

Over there on the left hand side of the dock is the newly built bridge of the aircraft carrier, all wonky diagonals and pixellated angles, and round the front of the big shed the carrier suddenly rears up in front of us.

'Are you OK with heights?' asks Derek.

Yes, I say.

'That's lucky,' he says, 'because there's a lot of them.'

We climb and climb, rising up the rigging, and with each step we ascend, more and more of Glasgow becomes visible. Up on the top deck of Lower Block 3 (or LB3) itself, I can see the whole skyline of the city with the Clyde directly beneath. Think of a twentieth-century British ship, any ship you've ever heard of except for the ones that sank, and it was almost certainly built along this stretch of water. Looking at the river now, the idea seems ridiculous. This river is tiny – a water feature surely designed by estate agents to add value to riverside properties. You couldn't fit a canoe on this thing, never mind the *Queen Mary* or the *QE2* or this section of supercarrier standing on the bankside sliced like a steel cutlet. Still, according to the plan, once each section is completed it's jacked up onto a transporter, inched out of the shed, loaded onto another ship, pulled down the river and transported round the whole coast of Scotland. Once it's reached the Firth of Forth it's then towed up the estuary and under the bridges, and finally joined together into a finished ship at Rosyth. By land, that journey is about forty miles. By sea, it's 600. The UK doesn't have a shipyard large enough to build this ship in one piece.

Directly above us is the roof of the shed. Below our feet is the top deck, and nailed to that deck are a series of small temporary huts for welding or storage. Behind a milk-white veil of polythene, a shadow-man stands spray-painting components, and over there beside the spools of coloured cabling the electrics are being sorted into strands. Beyond that in another cabin, three men are picking over drawings. Looking over at the tiny gap between the hull and the walls of the shed it's just about possible to imagine this as something that might one day move, but down below in the cabins full of sockets and pipework it just feels like the seventh floor of a rather gloomy office block. When it's finished, this ship will be a seaborne city, with doctors, dentists, admin staff, swimming pools, four separate canteens, plus of course a lot of kit for killing people. No wonder BAE have produced a mobile app to help people who get lost in the ship's innards. Are they really worried about people disappearing for good?

'Oh God, aye!' says one of the laser cutters cheerfully.

So has anyone needed the app yet?

'Maybe,' says McCaffrey. 'But they haven't admitted it.'

As the four of us return slowly to earth down the scaffolding, he looks back up at the bulk behind him. 'That's my baby,' he says.

Will he miss it once it's gone?

'I always miss ships when they go, yes.' The finished aircraft carriers are too big for a 'dynamic launch', the old-fashioned champagne baptism down the slip, 'but she does get named'.

So, despite Lloyd's 2003 announcement that from now on ships should no longer be considered feminine, does he still think of the carrier as she?

'A ship is always a she. What do Lloyd's know? We build her. She belongs to us.'

We look up at it again, craning our necks so far back that the tips of our hard hats brush the collars of our overalls. This ship has already taken months and months, thousands of hours, millions – billions – of pounds, all the uncountable thoughts, dreams, ideas, arguments, ambitions, revisions and speculations poured into it. It doesn't matter what it's called or how it's made, it's still the sum total of all the human energy that already belongs to it.

Back in the spring sunshine, we walk over to another building. I'm still overawed by the size of LB3, but the fabrication shed impresses in a completely different way. In through a little door in the wall and suddenly you're in a cathedral, a great secular cathedral designed for the worship of steel. The roof is way up in the dimness above and the light from the riverside streams in sideways, illuminating asymmetric slabs of metal laid out between taped lines along the floor. All the ground space has been divided into little rectangular fiefdoms – shot-blasting there, laser cutting here, a panel line for bulkhead assembly three miles to the left. Huge yellow gantries lift and reposition steel plates and over in one corner there's a giant flat bath filled with water over which a laser hovers, ready to razor out bits of ship accurate to within millimetres. To the south a group of men are welding T-bars and brackets onto some of the steel sections while over in the far west territories Jeremy Vine is talking to himself on Radio 2. Somewhere over there, about half a mile away there's a mocked-up section of a Type 26 destroyer, lots of square sections of space within a vastly greater one.

And it's noisy. In acoustic terms this place is a perfect echo chamber, a space designed to sweep sound straight back to you. The roof feels about six miles high and the floor space is – entirely literally – big enough to house a battleship. Or two. It's April, but it's still bloody cold in here. Even the welders are wearing balaclavas or beanies under their hard hats and the chill only amplifies things further. All the things which might absorb noise (wood, fabric, plastics) are absent, which means that there's nothing to fill the air except pigeons and that every wave returns for a second sounding. The walls of the shed are steel, the plates of the carrier are steel, the machines with which the plates are cut are steel. The floor is concrete and there's a river outside. Everything, every surface, every object, every material, is reverberant. Somewhere a long way away someone drops a wrench and as it clangs you can hear the sound flash out, hit the sides and ride down the building. It makes voices seem mousy and foolish and it means that people don't talk much until they're in a place in which they feel themselves back at the right scale again.

It might seem loud now, but in reality this place is a mere whisper of what it used to be. Partly that's because this yard is working at a fraction of its potential capacity. Partly it's because the way ships are constructed has changed. And partly it's because of health & safety. Within each metal fiefdom there are now bright HSE notices directing staff towards good behaviour, colour-coded notices insisting on

your attention to RIDDOR or COSH or heavy-lifting legislation. Everyone's wearing the proper gear – overalls, hi-vis
tabards, hard hats, goggles, welding gloves, workboots,
breathing apparatus or dust masks. And almost everyone
in here, whether they're welding or drilling or chalking up
plates, is wearing ear defenders.

It wasn't like this in the past. When shipbuilding first
took off here in the late nineteenth century, the yards were
so constantly, overwhelmingly noisy that workers developed
a basic form of sign language between themselves in order
to communicate at all. Within some trades, deafness was so
ubiquitous that it was taken as proof of experience. Just as no
one quite trusted a riveter with a full hand of fingers, so no
one would dream of hiring a forge operator who could actually hear. In 1886, a Glasgow surgeon named Thomas Barr
examined a group of 100 shipyard boilermakers and found
that 'not one of them had normal hearing'. Three-quarters of
the group were deafened to the extent that they were either
partially or entirely unable to hear someone speaking at a
public meeting. Barr found equal levels of deafness among
locksmiths, iron-turners, railway workers and weavers. Their
employers, the shipyard owners and industrialists, weren't
much bothered. Since deafness didn't stop workers doing
their jobs, deafness wasn't a problem. The prevailing attitude
then was that a worker could have three tongues, seventeen
toes and not a single sense remaining, just so long as none of
those things damaged their economic viability.

Even with Barr's work and the beginnings of legislation
offering some protection to workers' rights, hearing loss
remained more of an anecdotal problem than a legal issue.
Until well into the 1970s it was widely known that some pro-
fessions suffered more from hearing loss than others and that
some occupations – particularly those within heavy industry
– carried a much greater risk. The supporting science began
to build up and further studies began to isolate the exact

point at which repeated exposure to loud noise would begin to damage human hearing.

But still, it took a long, long time for anything to happen. When V. S. Pritchett visited several British shipyards working at their wartime peak, things were, if anything, even worse than when Barr had made his investigations. In a pamphlet written for the Ministry of Information and published after the war's end in 1946, Pritchett made it clear how overwhelming the shipyards could be:

> The noise of the [ship]building reaches a note and volume which are unimaginable. From a distance it sounds like a thick gale of wind in a forest; in the yard itself, as the rivetter's sparks dribble down from the ship's side, you seem to have got into the hot corner of a gunman's skirmish. In the yard you could hear if you shouted. Here, your shouts are knocked clean out. You have to dodge around a corner and hope one word in six reaches the ear that is leaned towards you. The roar comes from above, below and on either side of you, a pandemonium of clangings, rappings and sawn-off gun-work with men making rival roars in an alleyway a yard wide that at first causes terror as you grope through the darkness. Hundreds of men seem to be lying, kneeling, crouching, crawling about ... Once in a while, a face which has gone beyond indignation and resignation into a world of its own looks up from the level of your knees ... You look down in to the body of the ship through the smoke haze of the rivetters' fires and watch men step about there like little demons in the galleries of Dante's Hell.

Then as now, a warship was a complex organism and there might be 13,000 steel plates in its making. Beside the forge, 'Each man,' wrote Pritchett, 'stands by his hooded fire, his face smoky and reddened by flame and glistening with sweat.' He sees the concentration of the men near the forge, and watches their snake-killing dance as they lift the red-hot framing strips out of the fire and drive them with dogs and clamps to shift them from straights into S-bends. He watches the woman working the crane communicating with hand signals: one finger for stop, and two for go. And he notes the particular quality of sound in a ship: 'The rivetters' and caulkers' fusillade, the platers' solemn clang, and the elephantine thumping of the forge. These steel hammers that come down like tree trunks on the anvil shake the earth and the building and thicken the air with a cloud of reverberations.'

Welding was introduced in most British yards during the 1940s, but for the moment each plate would be joined to another with rivets – a million or more in the making of a ship, and all of them forged on site,

> The riveter is a member of the 'black squad', a gang of four who turn up to the job with the misleading nonchalance of a family of jugglers. [The black squads] can set up shop anywhere and begin performing their hot-chestnut act. You see one swung over the ships' side. He stands on his plank waiting with the pneumatic instrument in his gloved hands. On the other side of the plate, inside the ship, is the heater with his smoking brazier … he plucks a rivet out of the fire with his tongs, a 'boy' (nowadays it is often a girl in dungarees) catches the rivet in another pair of tongs and steps quickly with it to the holder up who puts it through the proper holes at the junction of the plates. As the pink nub of the rivet comes through, the pneumatic striker comes down on it, roaring out blows at the rate of about 700 hits a minute, and squeezes it flat.

An expert black squad, he notes, could manage about thirty-seven rivets an hour and there would be several squads working their way down the length of the hull. Since those men were all paid piecework, there was every incentive for all of them to bang away as hard and as fast as possible.

The noise – and the pain of that noise – was indescribable. Even now, with forty-odd years of HSE legislation, so much steel in such a great space is still loud. But back in the forties and fifties, for ten or more hours a day six days a week, the boilermakers, the riveters and the black squads existed in a world without quiet, perpetually exposed to between 120 and 150 decibels. They lived within a state of extremity – extreme scale, extreme heat, extreme cold, extreme noise, every sense perpetually overwhelmed:

> The individual is least, the group alone seems to have personality [wrote Pritchett]. And then, the sounds: this great and diabolical religion ... great and sudden clangs, an intoned mutter ... bell-like, gong-like crashes which astound the ear and the mind. You feel you may be watching a rite devoted to the creation of the ship which belongs naturally – before anyone else – to those votaries who are building her.

Back in London and writing up his experiences, he noted, 'It is the silence of the people in the noise of these yards which you think of afterwards.'

But, as the machines slowed to a stop on heavy industry during the second half of the twentieth century, the yards reverted to other sounds; the surrounding city, the river. In 1984, almost a century after Barr's identification of the damage caused to boilermakers' hearing, a landmark case finally established that employers could be held liable for that damage. Once the Noise at Work Act was passed, the responsibility for protecting workers' hearing passed to management. As understanding of the issue increased, so health and safety spread, and over the past 40 years, acoustics has become a standardised aspect of new construction contracts. In theory, we've now reached a point where no one should lose their hearing just for doing their job.

VICTOR HUMPHREY is a professor of engineering acoustics at the University of Southampton's Institute of Sound and Vibration with a particular speciality in research into marine and ultrasonic sound. He works in an old sixties block which he has filled with what appear to be the fruits of several decades' worth of study and lecturing – papers, books, photographs, drawings, models, instruments, things put there specifically and only in order to gather dust. He himself is tall and healthy looking, with thinning grey hair scraped forward and a habit of smiling faintly to himself when he says something technical. The smile is not meant to convey humour, but is a cross between a nervous gesture and a device for bridging the gap between the information and the listener. He starts off cautiously, punctuating his observations with thoughtful exhalations because – as he says – he's trying 'to get things absolutely right'. After about a quarter of an hour he becomes warmer and more enthusiastic and by the time I ask about the anechoic chamber, he's lit up like a small boy.

Human hearing, he points out, has been given a bad press. Compared to the fantastic whizz-bang capabilities of birds or whales, our ears have always been portrayed as a bit dim – no more than a detuned huddle of mid-range frequencies deaf to the great ultrasonic symphonies playing all around us. But, as Humphrey points out, human hearing is no less miraculous than that of a dog or a bat. Our bandwidth (our frequency range) may be narrow, but our capacity for amplitude is not.

Think of sound as a wave again. Think of that wave

becoming the power (energy) required to make the eardrum vibrate. 'We get our students to measure the amount of power that's radiated by a loudspeaker. So we put a loud-speaker onto the table, switch some noise onto it and ask them how much power is being radiated. They'll say things like one watt or ten watts. And we then measure that power. So if it were a light bulb and you put a hundred watts of power into it, you don't get a hundred watts of light out of it. Same for the loudspeaker – how much acoustic power is coming out? The acoustic power is .00001 of a watt. So the sound power is very small – very little energy. That's what's so amazing about hearing. It's just incredible, the hearing system, when you work out how small a pressure and what range of pressures you can detect.'

So the fact that we can hear birdsong through a window is really extraordinary? 'Yes. It's amazing the range of things we can hear, from the quietest thing to the very loudest, the sound that gives pain. That range is about a million in pres-sure variation. The quietest thing we can hear is a million times smaller than the things which would cause us pain.'

To illustrate his point he takes me out of the building through what seems like several other buildings, up stairs and down corridors to a big ground-floor passageway. Halfway along it is a big, high-ceilinged chamber covered in long wedge-shaped foam tentacles which reach all the way up the walls and hang down from the ceiling. In some parts of the room they stick out with such geometric regularity they create a sense of 3D illusion, but in others they've begun to droop, which makes the whole place seem flabby and worn. Below our feet is a metal grille laid in squares, and below that are more tentacles. Humphrey does not close the door, but even standing in the unsealed room is enough to give a sense of a place in which almost all sound has been muffled, as if each word had been coated in velvet. When he claps his hands, the sound is small and disappears immediately.

Despite the cold, I like the room – it makes speech and understanding seem effortless, though when I play it back the space translates as something tiny, as if Humphrey and I were talking to each other in a shoebox.

As he explains, the point of the room is that it has no resonance and so no echo. The foam on the walls stops the bounce of a wave, thus the only sound coming to us is coming direct from the source. In here, there are no sound waves pinging off flat surfaces so if someone standing to my right says something then I'm only going to hear them in the right ear, not the left. Mono, not stereo. It's exactly the effect that sound engineers try and create in a radio booth: fabric and absorbent baffles on the walls, small space, nothing to dissipate the single-source strength of a voice.

Humphrey takes me out of the chamber and into the room next door. This is the anechoic opposite, a resonance chamber. The room is large, rectangular, with a very high ceiling and wooden baffles hanging down at intervals. There is no absorption in here, nothing to receive the echoes, so all the noises we make – our footsteps, our voices – return indefinitely. There is so much reverberance in there it's like listening to the distant sling of tube trains approaching down the rails. Or a Glasgow shipyard. When Humphrey walks a little farther away, the reverberance becomes so great that it becomes difficult to pick up what he's saying. I have to watch his face much more carefully to hear him than in the anechoic chamber, and when he claps it sounds like something huge and flat dropped on the floor. The sound remains for minutes afterwards, making the space feel far bigger than it is, as if the size of the room is an illusion and we're both now lost in a mess of echoes. You would be able to hear a tiny sound in here, he points out, because even if your ears hadn't picked up the original source the reverberance fetches it back to you.

As Humphrey explains, the resonance chamber is designed partly to allow students to measure the power of sound from a particular source. So, for instance, they could put someone in here with a snare drum and then measure the pressure waves given off at different volumes. But because the sound in here is so bouncy, it's very difficult for the ears and thus the brain to sort out the source from the echo.

'I have to be standing closer to you in order for your hearing to recognise me as the dominant sound source rather than just one of the different reverberances,' he says. 'I'm trying deliberately to talk slower in here, but if you can imagine having a sound source in here as well, it really is quite difficult. And these baffles are designed to make that sound even more diffuse.'

It's interesting, I say, that the two rooms have such a strong psychological effect. The temperature in both rooms is exactly the same, but this one seems colder. Standing in here I just want to be next door in the anechoic chamber where the sound is so much softer. This place is like ... well, it's like half of Britain's bars.

Humphrey agrees. 'The problem with restaurants is that the designers like smooth, flat, shiny surfaces.'

Is that because they want to create the illusion that there're lots of people?

'Yes, but in general the problem is that it makes it difficult to have a real conversation. Absorption in room acoustics is what fraction of sound is reflected back from the walls. With a flat concrete wall like this, most of the sound would be reflected back so the absorption would be very small.'

As Humphrey points out, the irony is that what has now been outlawed as dangerous in the workplace is now actively encouraged outside of it. Great care and forethought has gone into making bars and restaurants as acoustically uncomfortable as possible. If three or four couples are talking to each other in a room full of curtains and soft chairs, they vanish. Put exactly the same eight people in a room with brick walls, high ceilings and a sound system, and suddenly they've become sixteen people, or twenty, or forty. From a restaurant's point of view they've just generated treble the covers for none of the price. Suddenly, this place – this empty-ish tapas bar, this Sunday Thai – sounds like something happening. So now there's a buzz about the place and that buzz is drawing in more people who have come to see what all the buzz is about. And in order to be heard above the rising volume, everyone has to lean in close to each other. The restaurant has created a space in which everyone is having trouble hearing, so everyone's begun to amplify their gestures and stare at each other's faces. Now they're not thinking about work and babysitters any more, they're thinking about sex. Plus, of course, the shouting has made them thirsty, so they're ordering more drinks. It's clever, really, until you realise that the reason for eating out is not the ambience or the buzz or even the food, it's the people you sit down with.

I look up and point to the ceiling of the resonance chamber. Does it make a difference that it's high? Humphrey nods. 'It has an impact on reverberation time, the volume of the room. And the walls in here are not parallel and the ceiling is not flat.' Why does that make a difference? 'If you have parallel walls, you tend to get some modes where the sound is just bouncing backwards and forwards and it builds up a standing wave in that direction. With walls like this, you don't get that and the sound is more uniform from place to place.'

He leads us out of the resonance chamber and into another area, in which a tiny booth has been constructed out of plywood boarding. The room has four amps – two at knee level and two at head height – and a series of small microphones poking out of the ceiling at regular intervals. Outside the booth two Korean students are tinkering with several mike stands to which are attached a cluster of loudspeakers.

When Humphrey asks them what they're doing, they say that they're trying to construct a 22-channel sound system, a 'virtual soundfield'.

What do you want to do with it? I ask. Home entertainment, they say. So the aim is to sit in your flat and watch *Spartacus: War of the Damned* as if half the Damned were in the room with you? The two students nod enthusiastically. 'That is correct,' one of them says, 'It will be a more realistic feeling.'

In fact, points out Humphrey, they could go one better and have completely different sound sources for each speaker, so if they stood in one corner of the room they'd hear a male voice and if they stood in another they'd hear a female voice, thus allowing them to get to know the Damned on an individual first-name basis. Both students look at him like they've just had some kind of profound spiritual experience. 'Yes,' says one of them eventually. '*Yes.*'

He leaves them to it and leads on. At first, the final room he shows me seems to have little to do with acoustics. This is the Human Factors research lab, a double-height space with a walkway at first-floor level and a deep pit at its centre. In the pit is a hydraulic lift topped with a stage, and on the stage is the chassis of a car. The lift can move on six different axes and its job is to shake like a road. Should you require it to, it can be programmed to simulate, say, a ten-minute ride through the New Forest or a five-minute drive across potholed city backstreets. Just above us is a TV screen on which is playing a vehicle motion test on seat suspension – a man strapped into half a car – and behind him is a computer simulation of a pre-loaded road journey. The car is stationery, but it is twitching in sync with the imaginary road.

On the right-hand side of the room there's a separate free-standing walkway which is designed to measure people's ability to function on board a vessel tilted at various angles, and elsewhere in the room they're testing machine tools like pneumatic drills or domestic sanders. Acoustics and vibration – and the way they're experienced by the human body – are so closely related that there's no point in trying to separate them, so this place does everything from testing Formula 1 cars till the nuts rattle to streamlining members of the British Olympic cycling team for wind resistance.

Acoustic science is both very old and very young. The effects of air or water on sound have been anecdotally understood for a long time, but Southampton is filling out that understanding and then applying it directly. To make ships and submarines quieter, the military have invested heavily in research into underwater acoustics. In addition to all the other established medical applications such as dissolving kidney stones, ultrasonics are now used to treat some cancers or establish the state of a patient's bone health. Business has got in on the act – dentists and shopping centres now routinely spend money on creating soundscapes designed to lull punters into a more pliant state. Acoustics, in other words, has gone from a state of almost non-existence 50 years ago to a science whose reach ripples wider by the day.

All this new research going on at Southampton and elsewhere only deepens our understanding that sound is a phenomenon with a potentially infinite number of attributes. If, instead of thinking of it as just noise, we thought of it as pressure or saw it as curves in the shape of time and space, then perhaps it would be easier to grasp its potential.

'Geology,' says Red in the film *The Shawshank Redemption*, 'is the study of pressure and time. That's all it takes, really. Pressure, and time.' As for geology, so for sound. And as for sound, so for the people who experience it.

8

Silence

FOR ALL THE TIME I've been writing I've kept notes in a series of small paper-bound books. They aren't diaries, they're just jottings about whatever it is I'm working on at the moment: thoughts or questions, bits of research, ideas, phone numbers. Most of the articles and all of the books I've written over the past two decades appear in some form or another, interspersed with more immediate to-do stuff. Because it's mostly just a record of work it's not exactly personal but if a subject engages me then it always makes it into the books. Each one is the same type – black, plain-lined. Some are sea-stained, some open to spatters of flapjack crumbs, and in all of them the handwriting varies all the way from meticulous to loopy. Every few pages my pen wanders off to the motion of a train or the spine cracks to a website address whose relevance I've long since forgotten. Writing longhand is an old-fashioned method of retaining information but the books are useful and I like them. Somehow or another they've become as much a record of my working life as the final printed results.

By 2004 I'd been losing my hearing for six years. I know by then I'd begun to observe the particular quirks of deafness – the difference in the effect of various instruments or the way it was perceived. So if I go back to the relevant books, I'm sure I'll find some kind of written evidence.

Instead, in all of them, there is nothing. Not a note, not a conversation, not a cry. Total silence. According to me, I never went deaf. I did plenty of research, I recorded more and more information, I worked harder and harder and faster and faster, but not one single word of it had anything to do with the thing often uppermost in my mind.

I flick through the books from 1998 onwards and then I stick them back on the shelf. I know exactly why there's nothing there.

By 2003, the fight-or-flight desire to party fast had begun to transmute into something slower and deeper. I fought the deafness – and myself – for as long as I could. Then I just got sad, and all the other nightmares came to stay. Audio-logically speaking, life had settled into a rhythm. I used the digital hearing aids all the time, I got them serviced once in a while, and I got on with life. Or rather, I – and my hearing – continued down. Down with my demons. If I'd been able to get any kind of a sense of balance I might have been able to see that life was not the wuthering drama I'd written myself into. But I'd found a condition – or a condition had found me – which seemed perfectly customised to my vulnerabilities. Like everyone else, I didn't want to be lonely or isolated or ridiculous or stupid or unprofessional or irrelevant. But as far as I was concerned deafness made me all of those things, and worse.

Somewhere underneath everything was a bigger truth. Deafness was neither good nor bad. It was completely neutral. It was just itself. What was good or bad was the way I took it, and I took it hard. Deafness acted on me like strong liquor, drawing out deeper hurts and older issues. What had begun as a kind of medical novelty spun down into a darker drive downwards, a bitter fuck-you attempt to rid me of myself.

Through most of this period I was still writing. One book in particular – *The Wreckers* – required a lot of inter-views with people all over the country. So I'd trot off with my digital recorder to the Scilly Isles or Cromer or Lerwick and talk to ex-lifeboatmen or coastguards or Receivers of Wreck. I remember a trip to the isle of Barra in the Outer Hebrides in November or February or one of those months with too much weather and not enough light. I'd set up a few interviews beforehand but I also went round the island making connections and persuading people to talk to me. Most were reluctant. I stood in pubs buying drinks for men who already had the word 'No' written all over them, or hunched on dripping doorsteps watching the blinds move aside, pause, and then swing emphatically back into place. Between searches I walked the Atlantic beach while the whirling sand smacked the skin off my face. Back at the deserted B&B in the evenings I thought, *I would go home, if there was anything to go home to.*

In films and books and newspapers, the world pivots on moments of action – the crash, the diagnosis, the unmasking – but in real life, that's not how it happens. Sometimes it's not the shock of the fight that does the damage, it's the accumulated silence in the time beforehand. Darkness works slowly.

It was the shame, that was the weight that lay over me. By then it seemed large enough to be a solid thing with a proper existence independent of me. When I woke up in the morning I could see it there at the side of my vision, and when I lay down to sleep I could feel the iron of it at the back of my breath. The shame blocked my throat. Every time I started to speak it stopped up the air. What could I say? What *was* there to say? And how could I convert something which seemed so much worse than words into something made out of ordinary English?

If you'd asked me at the time I would have said I was fine, absolutely fine. Really. Really fine. Honestly. Definitely. And if you'd pressed me I would have said that we'd done that topic now, so how about we move on and talk about you?

'What do you remember of me from that time?' I asked an old friend recently.

'Hostility,' he said. 'You were scary.'

'Scary?' I said, shocked.

'Yes,' he said. 'Closed off. If I asked you about how you were doing you'd just sidestep it, so our conversations felt completely one-sided. I talked about what was going on for me, but you never gave out more than just what you thought we wanted to hear. All of us could see that you were going through something, some kind of process or battle, but you'd never let us in. It sounds awful, but sometimes you were so far away it was just easier not to see you.'

The lower I got, the harder I worked. In just over a decade, I wrote six books. I taught myself photography, botany and the rudiments of marine engineering. I wrote countless articles, I sat in late-night libraries making notes for stillborn novels. I drove and I drove. I drove harder. I drove round and round the British Isles like someone chasing the laws of perpetual motion. And when I wasn't driving or typing or cycling or walking the dog in the park, I ran. I kept on running. *A moving target is harder to hit.* I wrote and wrote, phoned and typed and questioned and noted. I asked questions, I kept on asking questions, I asked and asked, but I never answered any questions myself. I wrote words by the thousand. I sat in a room typing, I typed and typed, I wore out whole keyboards in the pursuit of connection. Every minute of every day I was out there pinging away like Asdic. I saw friends and family as I always did, I ate and drank and slept and woke and laughed and loved and cried and did all the things that were supposed to prove me as a fully operational human being, but all the time it felt like someone – or something – had reached into my heart and, atom by atom, was drawing it out to its death.

It sounds daft and slightly melodramatic to say this now, but I know that for seven years or more there wasn't a single day when I didn't think about dying, and not many minutes in those days when I wasn't searching for the shape of it.

I knew absolutely that I wasn't going to take my own life, but I did often think longingly of more passive exits. I also made the discovery that there's more than one way to kill yourself. There's the active way, where you go out to seek death. You make the plans, you hoard the pills, you buy the rope. Or there's the passive way, where you just stand there on the threshold holding the door open, hoping – dreading – that death itself will take the first step towards you. I didn't actively want to die, but then nor did I actively want to live.

It makes me feel sad now, writing this. It was horrible. Or rather, it was frustrating being deaf, but it was horrible, really horrible, using the deafness as a judgement against myself. *Why call a friend? You're just hard work. Why keep writing? You're just making it up. Why try and be loved? You can't even be useful. Why live? Nobody needs you. Why die? Nobody's there.*

That shame was having an impact on other things as well. By late 1998 I'd split up with Euan. Over the next twelve years I went out with three men, though I can't claim that any of those relationships was very healthy. That wasn't the fault of the men concerned, it was just the inevitable consequence of my own anger and sadness. All of them were tolerant, generous and happy to adapt in whatever way it took. It was me who was convinced that hearing loss rendered me unfit for human consumption. If they asked me about the deafness, I'd be defensive. If they didn't, I'd be annoyed. On a bad day, they'd have to repeat everything they said at least five times. On a good day I'd just be knackered. Our social life together was often weird. Pubs were out of the question and parties were tricky. Dinners round at friends could be uncomfortable and at home we couldn't talk in the kitchen whilst doing anything kitchenish like cooking. If required to deal with two or more sounds – a phone conversation at the same time as frying something – I'd have to stop one and concentrate entirely on the other. It seemed I'd lost the capacity to multitask.

During the day I'd have the hearing aids in but at night I'd take them out and lay them on the bedside table. I'd be lying on my side just dropping off to sleep and through the thrum of his chest, I'd feel more than hear that my boyfriend had said something. Every time it would require a tiny calibration, a thought-out process over the potential of each statement. Did I sit up, put the aids back in and ask him to repeat? Or did I hope it was something requiring no acknowledgement? At night, things seemed so much starker than during the day. There were the things that could be said unspoken or there was a series of scheduled monologues conducted under the ochre glow of the street lights. 'If I'm being honest,' said Simon later, 'that really shocked me. I didn't really take in how deaf you were until I saw what it was like at night. When you took the aids out, there was nothing. You couldn't hear anything, and if you were facing the other way or couldn't feel me talking, then I might as well not have spoken. When I met you I knew you were deaf but the hearing aids gave the impression you were doing OK, so the difference was that much bigger.'

It was also unfortunate that things were at their worst at the moment when everyone else I knew was getting married and having children. I wanted the same thing, no question. I wanted to fall in love, have mad intemperate sex, stop having mad intemperate sex, start squabbling over door handles and savings rates, get lost, get found, say everything, say nothing. I wanted to know someone when they were scared or grieving or behaving badly. I wanted to go through crises with them, to love and fall out of love, forgive and be forgiven. I wanted to crack myself open, forgive my prejudices and run out of things to talk about in the car. I wanted the accumulated knowledge of years and days, I wanted to know someone other than myself *that* close. I wanted to stop being so selfish. Instead of this time-free, age-free, change-free rotation of years, I wanted the chronology of trimesters and half-term panics over childcare. I wanted to complain about being kicked from the inside, to go through the big messy adventure of kids and to take my place as part of the human succession. I wanted to be part of a family, my own family. I wanted the sort of knowledge that meant something.

But instead of knowing all that, I knew all the wrong things. I knew what Robert Louis Stevenson was doing in the summer of 1874 and how to take photographs using processes a hundred years out of date. I knew how to spoof a GPS set and the difference between flotsam, jetsam, ligan and wreck. I knew the correct procedure for dealing with Royal Fish and how to build my own bicycle. By the time I reached the other side of my thirties, I also knew the rust in my own voice when I hadn't spoken to another human being for three days or the feeling I got when I looked at the books I'd written and saw only a waste of paper. I knew hope, rotted. And the lower I got the more the probability of connection receded, and the more it receded, the more ashamed I got, and the more ashamed I got, the more I remained alone.

For a long time it didn't occur to me to ask for help because when you most need it, you're least able to ask for it. Sometimes when friends discuss what's really bothering them, they apologise: 'I'm sorry, I shouldn't go on like this, I don't know why I'm talking about myself.' But I did the exact same thing: talk about anything, anything at all, except the thing which is killing you.

Eventually it came down to a simple, binary choice. Change, or die. That choice didn't appear in some single cataclysmic event: no gunshots, no ODs, no sudden technical malfunctions. There was no moment of revelation, only the advancing knowledge that death had now metastasised to fill the whole of life.

But I had no real idea of what to do or where to look for help. I was deaf and I was sad, but what was I supposed to do with that? There was nobody out there to talk to about hearing loss because hearing loss was a joke. I couldn't confide in my friends because they were all knee deep in teething crises and plastic toys. I didn't need pills because I wasn't ill. And, I reasoned, there was no point in going to a therapist because I wasn't depressed, I just kept thinking about death.

And, because I'm here now to look back on it, I must have got out. I did get out. How I did so doesn't really matter – for everyone who doesn't die, the journey is totally different and exactly the same – though there were certainly a few tragicomic moments. I rang the Samaritans, but they spoke so quietly I couldn't hear them. I looked for some kind of gathering, some fantasy 12-Step group for people who weren't actually addicted to anything. I went to an AA meeting just for the fun of it. What I wanted was a group, a structure, peers – someone who could say, yes, I know who you are, and I know what you're coming from.

And finally, one Friday night, I turned up at the local mental health unit's inpatient department. I sat in the waiting room while the receptionist fetched someone from upstairs. One of the senior ward nurses came down and looked at me from the other side of the room. Then she came and sat down beside me.

Um, please, I said, I'm really sorry to bother you, but can you admit me?

'No,' she said.

Oh, I said. OK. Can you section me?

'No,' she said.

Please? I said.

'You do understand,' she said, 'what a Sectioning Order is, don't you? It's a legal document, and it's only used with someone who presents a risk to themselves or others and who won't consent to assessment or treatment. It needs several separate signatures, so it takes time to arrange.

Which therefore means we definitely can't use it for someone who turns up at the door at ten p.m. on a Friday night and begs to be admitted.'

OK, I said, I totally take your point, but could you section me anyway?

'Sorry,' she said. 'But no.'

Please, I said. Not even if I act unwilling?

'No,' she said, and smiled regretfully. 'Come back Monday.'

9

Distortion

SOME RELIGIONS HAVE A RITUAL which is only undertaken in the most extreme of circumstances. If a member of a family or of the community is considered to have so offended against it that an ordinary punishment is not considered enough, then an anathema is pronounced and that person is declared dead. A full funeral ceremony is conducted in their absence, the loss is mourned, and from that moment on they become a non-living entity, an ex-person.

Back then, I kept thinking about that ritual. What must it be like for that person, the anathematised, to see the very fact of themselves disowned? Who would do that to someone, and what could anyone have done that was terrible enough to deserve it? The trouble was, I already knew. There I was, in the middle of London, banished. And in this self-imposed exile there was no contact with the outside world – or no meaningful contact, anyway. There was no sustenance or softness, only a wintry survival.

At the time I thought that the best way of dealing with it all was just to pretend it didn't hurt. Physically speaking, that was true – hearing loss didn't hurt at all. Emotionally speaking, it was a different matter. Even so, I still believed I could get through this with nothing but coffee and strong will. It took me a phenomenally long time to realise that the things I thought were helping me to survive were the same things that were pulling me down, but I certainly wasn't going to drop any of them until I was backed so far up against the wall I had no choice but to put my hands up and surrender. Everything that I had held to – the strength, the endurance, the belief that I was alone – turned out to be wrong. Blissfully wrong. 'Why didn't you ask before?' said my friends, as soon as I did. 'I don't know,' I said, light-headed. 'I really don't know.' 'Come here,' they said. 'And give us a hug.'

After that, things became a lot easier. There's something very powerful in that game-over moment and even if I did blunder around for a while trying to figure out how to find the right sort of help, at least I'd asked for it and at least I finally knew I needed it. Perhaps the biggest thing was to stop the habit of hurling myself towards the things that most scared me, and to cure myself of the belief that if I just stopped being lazy I could hear as well as the rest of the world. I had found people who gave me the tools I needed to reconstruct myself. If darkness worked slowly then love, it turned out, worked at the speed of light. And, once I started working for a photographic lab, I found one of the keys to dealing with deafness hiding there in plain sight.

More than a decade later when I started researching sound and its effect on the military I kept meeting people who reminded me of that time. Up until then I suppose I'd been aware that soldiering required a lot of exposure to potentially deafening weaponry, but I hadn't realised the depth and extent of hearing loss or its wider psychological impact. These men had fought and won and bore the scars, and now all the things they'd fought for – the chance to have an ordinary family life or to sit peacefully at the head of their table – had been granted, but with the power of connection withdrawn. They could be in the room, they could see their children's futures taking place in front of them, but they couldn't make contact and they couldn't radio for assistance. They had been banished. Their past might have been explosive, but oh my God, how much worse was their present.

They were also terrible at asking for help – trained not to ask for it, in fact. That was what Queen and Country had told them to do: just shut up and take it.

I knew I hadn't gone through anything like their experience, so it seemed a bit presumptuous to superimpose anything of my own understanding. But somewhere through all those doing-fines and didn't-hurts, I always seemed to hear the echo of a brutal grief.

FROM THE MOMENT in the seventeenth century when gunpowder was introduced, hearing loss in the military has been a problem. Medical records from the American Civil War show that around a third of soldiers were deafened, two world wars made the issue bigger, and the introduction of the jet engine in the 1950s turned it into something approaching an epidemic. Guns and gunpowder are loud, but bombs, mortars, missiles, tanks, Typhoons, Vulcans and Harriers are sensationally loud.

Hearing's 'pain threshold' is around 120 dBs and a sound pressure wave of 150 dBs or above has the potential to burst the eardrum. If you were to stand with naked ears close to, let's say, a Eurofighter Typhoon preparing for take-off, it wouldn't just be loud to the point of excruciation but potentially loud enough to cause damage to the bones of the middle ear. Sound is pressure, which means the higher the volume, the greater the impact of that pressure on the body. An armoured personnel carrier can hit 110 dBs, a mortar fired at close quarters is about 190 dBs, the round of an M16 assault rifle firing is around 158 dBs, and the flight deck of an operational aircraft carrier can reach 125 dBs. No one, in other words, has ever invented a quiet way to blow things up.

The irony, of course, is that soldiering often requires great acuity of hearing. Survival may occasionally depend on being able to hear the click of a bolt or identify the alarm call of a bird. You need to be able to hear the enemy before the enemy hears you. But when you're not running around in jungles playing hide-and-seek with people doing their best to kill you quietly, you're in places where they're trying to do it as noisily as possible. When things do kick off and suddenly there's guns and bullets and explosions and noise so loud it does things to your vision, it's also vital to be able to hear the sudden exclamation from the person next to you because if you can't respond to an order, then there's a good chance that you're not going to live for very long. So you're expected to be in situations where your hearing is tuned to an almost musical level of discrimination, and then to be able to withstand the full bone-rattling impact of war. In both situations, conventional hearing protection isn't merely useless, it may well be actively counterproductive. Which means that, for as long as war has been loud, the army's choice has been either ear defenders or self-defence. Deaf, or dead: the choice is yours.

For a long time, the issue was either ignored or understated. Deafness was regarded as part of the job – an occupational hazard like PTSD or the clap. You joined the army, you got your hearing shot off. Besides, of all the possible injuries or disabilities caused by military service, NIHL (noise-induced hearing loss) seemed relatively mild. It wasn't like losing a limb. There wasn't blood, or grief, or whatever horrible thing happened to your insides when your best mate died beside you. Hearing just snuck off silently in the middle of the night or exchanged itself for tinnitus's empty fizz. Nobody saw it happen, nobody mourned its loss. Troubling, but on the one-to-ten scale of military horror, pretty low on the list.

And, despite the evident correlation between loud bangs and hearing loss, the armed forces remained reluctant to acknowledge the issue. Partly that's because deafness and tinnitus are the two most prevalent service-related disabilities, and thus the potential implications of large-scale NIHL compensation claims are too terrible to contemplate. Even at current levels, disability payments to veterans in the USA for hearing loss and tinnitus are well over $1 billion a year. What it would be like if the true number of deafened veterans came forward to claim, nobody knew, and nobody wanted to know. By the Ministry of Defence's own estimate, things were not much better in Britain. Up to February 2016, 12,622 claims for NIHL compensation had been brought against the MoD, 9,388 of which were successful. A 2014 response to a Freedom of Information request said that of the 156,220 personnel serving at the time, 3,980 had been diagnosed with some form of hearing loss, though the same response admits that the figures represent at best 'a minimum burden of ill-health'. For 756 of those individuals, their hearing loss was severe enough to warrant their discharge from the service.

The MoD's assessment is, in all probability, a radical underestimate. A 2008 study in the USA suggests that more than half – 51.8 per cent – of personnel still serving at the time had moderate to severe hearing loss. During the first year of the war in Iraq, 'an average of one soldier a day was medically evacuated for complaints related to hearing loss'. That's a lot of soldiers, and a lot to lose.

But you have to look quite hard to find the long-term emotional consequences of all that loss. The military chat forums offer plenty of conversations about the practicalities of audiological testing or army pension arrangements: who pays what for how much loss, etc. But underneath that runs a quieter hum of private pain. Front-line soldiers worrying about being downgraded to support roles, older vets beached in zero-hours civilian jobs wondering if they can retrospectively claim for compensation years after they left. 'I refuse to wear hearing aids at work as it's embarrassing' ... 'Damned ear defenders don't defend much when putting a few dozen rounds through the Scorpion's 76mm' … 'Unless you can blag another hearing test with a "friendly" medic, your fooked [sic]'.

Nobody likes it. Nobody wants to admit how much they don't like it. And, just as with PTSD, nobody wants to point out that the qualities needed for a job in the armed forces are the reverse of the qualities needed to cope best with deafness. The whole point about being in the military is that you're supposed to be fierce, strong, a dutiful team player, someone who's capable of putting aside their own emotions and individuality for the sake of the greater whole. So how's that compatible with someone who has to ask five times for the way to the toilets?

The trouble is that there's collusion on both sides. If, let's say, you're training to be a fighter pilot and you're one of the fortunate few who has spent the four years and £4 million necessary to complete that training and you then wake up one day with the absolute knowledge that you're going deaf, what then? Annual medical tests were only made compulsory for all British service personnel in 2008, and even now hearing tests aren't conducted by a specialist. In the past it was up to the individual themselves to have a long hard chat with their own conscience and make the decision to report the problem. Once they'd done so, and depending on the extent of their hearing loss, there was a strong chance that they would either be withdrawn from service completely or demoted to a non-flying role. Or they could make the decision to keep quiet, keep the job, and risk the lives of everyone around them by mishearing a command or blanking a crucial auditory clue.

From the forces' point of view there was also a vested interest in making the best of each case, not just because of the compensation issue but because of the resources they'd already invested in each individual. Though the MoD is bound by the same Noise at Work regulations as every other employer in the UK, they also have trouble recruiting and keeping staff, which means they can always theoretically argue that getting rid of trained individuals just because they're having a spot of trouble with high frequencies is a very inefficient use of tax-payers' money. Thus army doctors had several good reasons for downplaying a hearing loss diagnosis and soldiers were generally left to make the decision themselves.

All of which is really quite alarming. After all, if your commanding officer yells, 'Fire! Fire at the large scary thing over there!' and you hear nothing but a vague itching in your earphones, then either the scary thing fires back at you or you fire at the not-scarey thing in the other direction, thus potentially causing both a tragic meaningless death *and* a major geopolitical incident. It's not so improbable: a 2008 study in the USA found a significant difference between the response times among tank crewmen who hadn't heard a command properly and those who had. Among those who had been able to hear a command to engage a target, 94 per cent had hit the right one. Of those who hadn't been able to hear the command, 41 per cent had hit the right one. Which means 59 per cent hit the wrong object. Or person. Or people.

Attempts to improve the situation have produced mixed results. It wasn't until large numbers of service person-nel reported problems after both world wars that the MoD began considering the question of protection. If recruits were having trouble with their hearing, they suggested, then perhaps they could try cotton wool. Or maybe cotton wool moistened with Vaseline. Or sticking their fingers in their ears. When large numbers of soldiers in the Irish Army brought cases against the Department of Defence between 1999 and 2002, some reported that they had been forced to improvise their own ear plugs out of the filters on cigarettes.

During the war in Afghanistan the US army introduced the combat arms earplug which blocks loud-impact sounds but lets through soft fluctuating sound such as voices. In theory, all personnel were supposed to use it. In practice, it was shunned for being too expensive (at $6 a pair), and for the usual situational-awareness reasons.

In Britain, the Personal Interfaced Hearing Protection (PIHP) was introduced to all those serving in Afghanistan. When later surveyed it turned out that only 4 per cent of those who responded were using them. Another study found that, among flight deck crews, four-fifths were either not wearing earplugs or wearing them wrongly, and one double-sided model so puzzled its users they just chopped it in half, thus rendering it marginally less useful than nothing at all.

SOMEHOW GENERAL SIR PETER DE LA BILLIÈRE, KCB, KBE, DSO, MC, managed to be right there at all the crucial moments of the late twentieth century. Born in 1934, he endured a childhood which sounds truly grim – father killed during World War II, mother incapacitated by an accident, escaped from a fire at his prep school aged ten, joined the army as soon as legally possible. Having worked his way up the ranks in Korea, Borneo and Malaya, he was recruited to the SAS in 1956, presided over the Iranian embassy siege in 1980, stayed on to sort out the Falklands after the war, and was appointed Commander-in-Chief of British Forces during the first Gulf War. Opinions on his legacy are divided – some hold him up as the godfather of the modern SAS, others say his biggest victories lay in advancing his own cause – but either way, he has certainly stacked up an impressive range of medals and awards throughout his service. Military Cross in 1959 with an extra bar a few years later, Distinguished Service Order in 1976, Commander of the US's Legion of Merit in 1993. 'Not particularly bright,' said a high-ranking colleague who had served alongside him, 'but definitely the bravest man I've ever met.'

Sir Peter is just how you'd expect a very distinguished

army general to be: large, formal, faintly menacing, a deep, roary-drawly voice. His wife Bridget is friendly but shy, small in both height and speech, as if she's spent so long taking the full weight of Sir Peter's extremely strong personality on her shoulders that she has somehow got smaller along the way. For the interview we walk over to a separate building with better acoustics, an old piggery which has been converted into a games room and snug. 'Do you like it?' he asked. 'It's our grandchild trap.'

Sir Peter has been dealing with deafness for much of his life. He had always been colour blind, but by the time he was leading patrols in Aden in his late twenties it was clear that his hearing had begun to degenerate.

At the time, they were out in the jungle fighting an unseen enemy. 'And of course if you're on a patrol, you need to be able to hear the slightest sound before the person making it hears you. And you'd use light signals – or in those days you did: red and green lights fired from these things called Vari-Lites.' The signals were relatively simple – something along the lines of red for withdraw, green for advance, though since he couldn't discriminate between the two colours he was unable to interpret their message. He enlisted the services of a signaller called Geordie Low and between them they devised a set of simple codes, allowing Sir Peter to circumvent both the hearing problem and the colour blindness. 'The noise was the real issue. I had a system with him, and as far as I remember it now, it was one tap on the shoulder for a noise, two for red and three for green, or something like that, so that he didn't have to talk to me on patrol. You don't want to start having discussions if you're creeping around in the dark on the side of the mountain with a lot of other people creeping around against you.'

Back in Britain three years later in 1967, he was promoted to the rank of major and sent to staff college at Camberley. Taken out of the battlefield and into the tutorial room, Sir

Peter found himself encountering a different problem. '[The deafness] was damaging my ability to take in lectures, which was embarrassing because then I couldn't respond to them properly at the academic level that they expected. I couldn't hear or I misheard, which was in a way even more dangerous because you answer what you think you've heard, which isn't necessarily the answer that they are looking for. Anyway, I got so fed up with this it forced me to go and complain about it to see if I could get something done. Not out of a sense of duty but because I wanted a doctor to do something so I bloody well could hear. It was just irritating. And out of that came the fact that I was deaf – to the point where they downgraded me. I mean, I was thirty-six, that's the end of your career, really.'

Sir Peter appealed. 'I was in headquarters, I was running things, and I didn't need the fine-tuned hearing that I needed when I was younger, leading patrols or listening out for enemy movements at night in the jungle or in the desert or whatever. I argued that at that age it didn't matter, which to some extent it didn't because one was more into giving orders than receiving them. But of course it did matter.' Either way, he won the point. As he writes in his memoirs, 'My appeal was accepted, on condition that I would take a special test every three years ... and, whether by good luck or good management, I contrived to be abroad every time the date came round.'

After that Sir Peter continued to rise, first in Special Forces and then at the Special Air Service. He played a key role in the Iranian embassy siege of 1980 whilst serving as Director of the SAS. How was deafness affecting him at the time of the siege? 'I was in the Cabinet Office briefing and advising other senior politicians and police officers rather than at the scene of the incident. The people at the scene of the incident who were the ones who suffered noise exposure would have been [the rank of] major downwards. After the rank of lieutenant colonel, I was promoted out of deafness, if you see what I mean, because I would have been in more senior appointments, and therefore more managerial and farther back from the front line where the firing is.'

At that stage, was everyone of higher rank deaf? 'Yes. Yes.' So it was pervasive? 'Yes. Everybody in the army of my generation who went to war and fired weapons had some degree of deafness. Well, I say everybody, but a very high percentage. And bearing in mind that there was inadequate or no hearing protection – cotton wool or something like that – deafness was quite generally accepted in the latter stages of an army life.' Were there other senior officers in the same position as you? 'Yes.'

So how did he deal with things during the Falklands and the Gulf? 'Well, that's a good point because there would be very critical meetings and I would have to make decisions, requiring a full understanding of what was being said in order to come to the right decision which people's lives might depend on. And so once again you come back to two things: telling people who are sitting around talking with you that your hearing is not good and asking them to make sure you've heard, and – most important, really – having an officer or people with you who are aware of your disability and who will be listening out for you and who will notice from your behavioural pattern or from the answer that you give that you haven't heard something and put you right.'

Would the RAF or the army allow someone deafened to continue flying or operate a tank? 'The more senior you become, the more experienced you become, the less your level of hearing impairment will impact on your employability. So there's a balance coming into play – it's never all in your favour as a deaf person, but it mitigates a deteriorating situation.' Which is interesting, given that the implication is the more senior you are, the less you actually need to listen.

Sir Peter has been retired for many years. His hearing has continued to degenerate steadily, and he now also has tinnitus. After leaving the army he went first into banking, and then writing, starting with his own autobiography, *Looking for Trouble*. He found it; the book prompted a row over his (very discreet) revelations about a service famed for secrecy. Several other books followed, and for a time Sir Peter took on a new role as a public speaker. That too proved tricky. When speaking at book festivals or big events he would almost always be unable to hear questions from the audience, and took to having someone sitting next to him to relay questions directly to him.

'What I was terrified of was misunderstanding and giving the wrong answer. You see, the trouble is, it's not just a question of hearing what you say, I anticipate not hearing subconsciously. And so I'm more concerned about trying to hear you than about what you're asking me.' It's a feeling familiar to anyone who's losing their hearing; you end up so preoccupied with trying to grab a handful of words and stuff them down your ears that you forget about their meaning. You're working so hard to hear that you're unable to listen.'

As he points out, as hearing gradually slips away, there's a strong probability it takes your social life with it. First of all he stopped going to the theatre – 'I'd just go to sleep – there's no point, I can't hear, I'd miss far too much of it.' Then he found that he couldn't hear church services, then concerts became pointless, then he stopped going to dinner parties. Finally even family dinners became tricky. His grandchildren are still young, and, 'Oh, it's very annoying, it's one of the most annoying things of all, because I can't hear children's voices – either the pitch or the volume – and so I try to join in around the table or sit here listening to them, and find that I can't hear them. I keep asking them and I feel more embarrassed about being deaf in front of my grandchildren than I do amongst my contemporaries. They're very good and understanding about it now because they know about it, but it does mean I drop out of the conversation completely and I stop trying to listen in the end. So if there's a conversation going on at lunch around the table, I'll drop out.'

He changed his car to something much quieter, he sound-proofed the barn, he stopped answering the phone. 'So in the end you say, well, why bother to go out? It's all such an effort, it's embarrassing other people, I don't get much out of it, I'd rather stay at home. So you stay in the grandchild trap.' So the danger is that it becomes isolating? 'Yes. Exactly that, exactly that. It does. And the thing is, you don't realise. I've been very busy until about three years ago, and I suddenly realised, yes, it is isolating ... I'm tremendously reliant on Bridget, and she doesn't help really when she talks in a quiet voice and more often to the Aga than to me, but at least she isn't surprised when I haven't heard and keep asking for repetition.'

He probably wouldn't go to a restaurant now. 'Because I know I won't be able to hear what's being said, and it subconsciously makes one antisocial, sub-social. I don't want to move out of this house where I can hear everything and I know where I am. And so I don't want to go on holidays, so poor old Bridget doesn't get holidays unless she bullies me. And that's the way it impacts on me – it makes you a recluse.'

Does he have friends in a similar situation? 'Yes. Particularly service friends. Some quite seriously bad.' And is their experience similar? 'Yes, I think it probably is. I mean, you don't go out to talk deafness, and the chances are if you're able to talk deafness to somebody, then you know them pretty well anyway, so it isn't the sort of issue to talk about very much: "Yes, I'm deaf, yes, you're deaf; bloody nuisance, isn't it?" You know, and there's an end of it. They understand it, and you understand them.'

If someone gave him his hearing back tomorrow, what would it change? 'Well, I'd really enjoy being able to hear birds which I can't hear at all and which everybody else can

– they talk about them. A couple of days ago, Bridget said, "Oh, I heard the birds this morning." I'd love that. It would give me great enjoyment to be able to hear my children, and they would be delighted not to have to shout at me the whole time, and of course bring me into the conversation which they don't bother too much about because they know I can't hear them. I don't mean in a nasty way – "He can't hear, so we're talking to everyone else except him. And if he can hear, well, that's fine." It would take a lot of pressure out of life, I think. But you don't think about these things now because they're not possible, so why bother?'

OLIVER HEADLEY* is a former bomb disposal expert with the Royal Ordnance Corps who was later seconded to Special Forces and retired a decade ago. He'd been suggested as a good person to speak to, so I emailed him and arranged to meet in a Home Counties pub. He's heavyset with dark hair, wraparound glasses and a striped shirt with the collar turned up. His eyes aren't blank but they're not warm either. There's something there, but it's very far back. He speaks with a very particular inflection, throwing his statements down like a provocation, like the show of cards that ends the game – *There. Done. Bang* – though maybe it's just the way you talk when everyone expects you to have an answer even when you don't.

Like most soldiers he spent a lot of time in planes. In Chinooks, noise levels could reach 120 dBs, and though he and the rest of the crew would be wearing helmets fitted with intercoms (ICs), the helmets were designed to protect their heads, not look after their hearing. If they wanted hearing protection while they were on ground operations, Headley says, they bought their own.

Not that it got much use because it seemed like every time they needed it on, they had to take it off. 'You can't

---

* Oliver Headley is a pseudonym.

operate sometimes when you have hearing protection on. Especially when you're on live operations, say. I mean, it's far more important to communicate than to protect your hearing, especially when things are going ...' He pauses, searching for a polite form of words. ' ... Yeah. Pretty Pete Tong, you know? Because if you've got someone with double hearing protection, you can't communicate with them, so the only way to get his attention is to stick your rifle butt in his head and get his attention that way, or physically grab hold of him – this is not what you need to be doing. You need to be able to communicate, and that means a lot of screaming, a lot of shouting. So a lot of the time, we operated without any hearing protection whatsoever.'

Headley retired a decade ago. Does he think things are still the same?

'Yes. Absolutely. Because if you start screwing around doing that sort of thing, you're going to end up getting killed. Or badly wounded. Or the other guys are going to do that. So, you know, it's a trade-off – hearing suffering, or coming back intact? It's a no-brainer, really. And this is what happens, and this is what has always happened, and this is what will always happen in the future.'

Does the MoD understand that soldiers have a conflict between doing their job and protecting their hearing? 'Their attitude is just the same as it's always been. They're very unhelpful. They do not accept that, well, hang on, there's a necessity to communicate on operations. And they'll just weave their way out of it. They don't recognise tinnitus because you can't prove it. When I went to the veterans' agency about it, they said, "You can't prove tinnitus, so you can't claim." And that's the reality of it.' It's 'hysterical deafness' all over again.

How much do they accept people continuing to serve with severe hearing loss?

'If the medical officer finds that you should be justifiably downgraded, you are downgraded immediately, and you're suspended from operations like that.' He clicks his fingers. 'So a lot of the guys over the years, myself included, have covered up injuries because we didn't want to be removed from operations.'

How do you cover things up?

'It's not that difficult to bluff a hearing test if you know how to do it. Which I've done in the past. You bluff it by speaking to someone who's had the test and asking them, "OK, so in between each tone, what's the time gap?" "Oh, round about fifteen seconds." So every fifteen seconds, you just keep hitting the button.' He chuckles. 'Guys who are in the Special Forces are cunning buggers by definition, and if they can't find a way to bluff the hearing test, they shouldn't really be there.'

'So basically you've got a bunch of deaf …'

'I wouldn't say a bunch, no, you've got people who are adversely affected by things who are keeping things quiet or they are manipulating the system in order to continue on operations, yeah, sure.'

With the tacit consent of the MoD?

'We are expendable. Everyone is expendable.'

In his case the damage was partly caused by the noise from helicopters, and partly from thousands and thousands of rounds of ammunition let off at close quarters, and from explosives. 'A rifle bullet typically when it leaves the muzzle is travelling at say seven to eight hundred metres a second, right? High explosive, when it detonates, is detonating at 8,500 metres a second. There's a massive difference there, and it's extremely shocking to the hearing, and to the cavities of the body when you're in a room and it's going off.'

Immediately after you're exposed to that kind of shock, what do you feel?

'OK, well, a high-explosive shock wave attacks the cavi-ties of the body. So the cavities of the body are the thorax, the abdomen and the cranium, and the shock wave passes through it. And you feel it hit you as it goes through the cavities of the body. And when you're in a building, it can be very incapacitating. In fact we use it as a weapon. Special Forces would use high explosives to go through a wall, but they would use more than is necessary to breach the hole in order to use the over-pressure on the bad guys inside to incapacitate them before they go in.'

So they would be physically wiped out?

'It depends on the amount of explosive used. To take that door off' – he points to the flimsy pine door between the pub lounge and the bar – 'you don't need much explosive just to cut through it. What we'd call a single strand of det cord – detonating cord. But if that went bang, it would be terribly painful, terribly shocking for you, and you would feel the shock wave. But if there were bad guys on the other side of that door I would use maybe ten strands of det cord, which is massive overkill, but it would definitely bugger them up inside before we went through the door. And when you're doing this kind of stuff live on operations you don't use the formula that is mathematically proven to cut that door with explosives – you use the P Formula, which is P for Plenty, in order to get the desired effect on whoever is on the other side.'

Overkill, literally?

'Massive, yeah. And when you're breaching a wall, you use massive overkill before you go in, absolutely. And you are in close proximity to this, and the shock wave – I mean, you've been exposed to it for so long, you know what's coming, but it's still shocking every time it goes off.'

What does that do to you psychologically?

'Hmmff. Nothing.' An intake of breath. 'No, nothing at all. Psychologically, absolutely not. What, you mean using explosives?'

Yes – that kind of physical shock.

'Oh, you just get used to it.'

OK, what does it do to you the first time, then?

'Well, it's just a shock – it doesn't do anything to you psychologically. It's just, "OK, that was different." And you just get used to it. To the point where you don't even jump. When you're going through a building and you're blowing door after door after door day after day, week after week, month after month, you just become immune to it. You know it's going to go bang, you know the shock wave is going to hit you, yeah – but it's no problem, provided you're not too close to it and the explosives aren't big enough to really affect you.'

Silence. He takes a sip of water.

What are you now?

'I'm retired. I'm a fisherman and a hunter. That's what I am now.'

How does that time in the forces affect your present life?

'In lots of ways really.' He's reflective, quiet. 'I suppose you can't help being shaped and moulded into certain ways having spent so long in and operated at the levels I've operated at. And I suppose seeing what I've seen. But beyond that, I really don't want to get into it.'

OK. How do you feel about the military now?

'Oh, I wish I'd been a dentist, I really do.'

AS HEADLEY POINTS OUT, the military have now turned their enforced expertise in very loud noises into weapons in their own right. In addition to techniques such as waterboarding, which became notorious during the Iraq war, the US military still legally uses so-called 'white torture' as part of its arsenal of 'enhanced interrogation techniques'. Considered to be more 'humane' than traditional physical forms, no-touch techniques include the use of 'futility music' to confuse and disorientate a prisoner psychologically and to break down their resistance before interrogation.

The British used loud and continuous noise as one of

the 'five techniques' of interrogation against IRA suspects until the practice was ruled 'degrading and inhuman' by the European Court of Human Rights in 1976 and outlawed. Some time later the Americans picked up the five-techniques idea, using it first in Vietnam and then against the Panamanian dictator Manuel Noriega. But the use of futility music really came into its own during the Iraq war. Just as during the 1960s and 1970s the CIA had experimented with various stimulants and hallucinogens, so they now tried to do the same thing by so overloading or depriving a detainee's senses that their hold on reality loosened. And they discovered that the use of music – or its opposite, white noise – cuts straight to the core. Half the point of music is to override logic, and the other half is to draw out an emotional response. Expose Arabic detainees – some of whom are not used to Western music – to uber-American rock at extreme volume and the effect is, literally, torture.

In 2014, the US Senate Select Committee on Intelligence issued a report on the 'enhanced interrogation techniques' favoured by the CIA during the War on Terror. The report found that 'Conditions at CIA detention sites were poor, and were especially bleak early in the program. CIA detainees at the COBALT detention facility were kept in complete darkness and constantly shackled in isolated cells with loud noise or music and only a bucket to use for human waste.'

A British citizen, Shafiq Rasul, who was held in 2003 at Guantanamo Bay while suspected of affiliation to Al-Qaeda, would be left chained in his cell with music playing on a loop at extreme volume for twelve hours a day. In a 2006 interview for the magazine *SPIN*, he explained its effects. 'It just starts playing with you,' he says. 'Even if you were shouting, the music was too loud – nobody would be able to hear you. You're there for hours and hours, and they're constantly playing the same music. All that builds up. You start hallucinating.' A year later, he was released without charge, evidence having emerged that in 2000 he was not, as the authorities claimed, hanging out with Osama bin Laden plotting the 9/11 attacks, but at college in the UK with a part-time job at Curry's.

Because the playlist at Abu Ghraib and Guantanamo Bay apparently included not only heavy metal (Led Zeppelin, Rage Against the Machine, Metallica) but Britney Spears and Barney the Purple Dinosaur, the idea of using music to break the will of a detainee was somehow seen as funny. It wasn't. The long-term use of any sound above 85–90 dBs can damage hearing, whether it's birdsong or Black Sabbath. A prisoner's body might heal after physical torture but their mind probably won't, and, once damaged, their hearing would be gone for good.

Despite the protests of several groups and individuals whose music has been used in this way – Trent Reznor of Nine Inch Nails tried to take the US army to court – and the findings of the Senate Committee that no useful intelligence had resulted from the use of white noise, it remains a legal technique in the USA.

So what does Headley know about white torture?

'White noise? For interrogation? It's not extreme in any form. It's like a crackling, a static. Crrrrrr, like you used to get with old-fashioned wireless sets when you twiddled the dial and it went between radio channels. There would be the crackling, until you got on to the next channel. That's just what white noise is – it's just turned up, there's nothing extreme about it.'

What do you use it for?

A pause, a calibration.

'You would use it to psychologically wear people down in an interrogation setting, so they'd be blindfolded probably and maybe even in a stress position where they'd remain for days on end, and you'll expose them to the white noise. But it's not painful to the hearing.'

So it's not ramped up to a level which is physically painful?

'No. Never in my experience, no.'

And what's the effect of white noise over a prolonged period?

'It just wears them down. It's just a psychological tool to … you know, deprive someone of sleep, detach them from the surroundings that they perceive themselves to be in. It's to remove them from any form of comfort zone and to put them in a place where you want them to be psychologically to subject them to various interrogation techniques.'

And does it work?

'Um. Because of rules and regulations, you can't apply' – a chuckle – 'the sort of interrogation techniques that would be most effective. Rules and regulations stipulate that you're only allowed to carry out interrogation in a certain fashion and white noise is a part of that, but it's not cranked up to a point that is painful for the person who is in receipt of all this.'

And if rules and regulations didn't exist, it would be?

'If we could interrogate people in … um, the way that would be most effective, we would get a lot more out of them. And things would be a damn sight more uncomfortable for them.' He holds my gaze steady. Another silence.

So you would break them down quicker?

'You'd break them down a lot more quickly and you would get a lot more accurate information out of them in a shorter time-frame. Definitely. So when you hear about all this stuff that went on in Iraq and Afghanistan, the media have just hyped it all up. And as I say, it's the same with white noise or music or anything like that. It can be repetitive, it can be unpleasant or it can be damn painful if it's really cranked up, but in my experience it's never done like that.'

So you used music as well?

'Yeah, I mean … So if we had personnel – whoever they were – deploying, part of that pre-deployment training would be to prepare them for how to handle the situation if they were taken hostage. It's not Arabic music, if you like, it's extremist, insurgent's music which they would listen to, which is extremely religiously orientated if they're from their faith. So the Western students would be bombarded with this for days and nights on end. Constantly. Not white noise. Because the insurgents don't use white noise.'

So you use religious music?

'Yes. You use religious music. From their side – we don't use any genteel choir singing or anything like that, it would · be extremist stuff that the terrorists listen to.'

The discussion of it I've seen in the media is about the use of heavy metal, etc., against Muslim suspects because it comes from a different world view and because it's hammering an alien Western message …

'… that they hate anyway, and it just pisses them off. And it's done to get a reaction. Or it's done to wear them down. Sure.'

But it's not done at painful levels – the point is to be relentless, not to destroy hearing?

'In my experience, no. No.'

So you use sound as a weapon yourselves, plus you're exposed to it on a continual basis yourselves …

'It's not used as a weapon, it's used as a tool. Very different things.'

Headley's account is of a world in which all the good things in life – the tastes, sounds, sights, smells and sensations, the human connections – are twisted down into something dark. Listening to him, I'm torn between revulsion and sorrow. If he and the army have used cruelty against others, then surely they have used that cruelty threefold on themselves. It reminds me of his description of those pressure waves. ·

Bang! The first time you feel something. Then you feel nothing.

What do you know?

Nothing.

What does this mean?

Nothing.

What do you feel?

Nothing.

What do you feel?

Nothing. I feel nothing at all.

I don't believe you.

10

Sign

BY 2006 I WAS AS USED to the deafness as I was ever going to be. I'd developed ways of dealing with it, and ways of circumventing it. If I was walking down the street beside someone I'd automatically swap places so I was on the clearer side of them. If we were going anywhere in the car I'd be the passenger because I could hear better with the right ear. When I met someone new I'd learned that it was much easier to explain that I was partially deaf, and I was glad now when people felt they could ask questions about it. I didn't go sailing so often and I'd developed a fondness for foreign films with subtitles. I had laser eye surgery to correct short-sightedness, which meant that I could actually see what people's faces were saying.

I also signed up for a weekly evening course in Level One BSL. In the end I never completed it – work commitments got in the way – but I did learn some of the basics, plus extras besides. As usual, what seemed most interesting was not the content of the course but the other people on it. The tutor was prelingually deaf, but all the students were female, youngish and – apart from a few who had a family member with age-related hearing loss – had no particular connection to deafness. They were taking the course not because they were losing their hearing themselves, but because they were actors curious about the physicality of BSL, or because they were studying languages, or because they worked in some sector which brought them into contact with the Deaf.

Once we'd got past finger-spelling we moved on to individual signs. For a couple of weeks all of us struggled with this new sensation of conveying concepts through our bodies. To begin with it seemed so counter-intuitive, scrunching up our faces and silently mouthing words at the same time as shaping signs with our hands. Normal speech required standing stock still, but this was all about motion, balance, expression. Sometimes, getting our point across required touching the other person, sometimes we communicated with our eyes or through our breath. We laughed a lot, partly to cover our embarrassment, but mostly because we all felt a curious sense of release. Signing was like dancing; it had rhythm.

Signing between the Deaf has always existed. Both Socrates and St Augustine make passing reference to it, and in the seventeenth century a young doctor named Bulwer composed a complete lexicon of hand signs which he called a chirogram. He picked out 64 basic signs from which others could be composed, including the traditional gestures of supplication, blessing, applause, surrender and triumph.

A century later a young curate in Paris, the Abbé de l'Épée, became the legal guardian of two young deaf sisters. He decided that he could either allow them to remain as they were – which would restrict them to a life spent communicating only with each other – or he could try and find a way for the outside world to reach them. To begin with he tried the obvious route; pointing to objects (tree, table, glass) and then naming them in writing. Verbs were covered by the same method – if, for instance, he wanted to say 'take the glass upstairs', he had only to mime the various objects and actions, and the sisters would understand. His method worked fine until he got on to the subject of God. The Abbé's intention had always been to teach the girls about faith, although this was more difficult since faith was not a tidy noun. Somehow, he realised, he was going to have to find a way of physically expressing an abstract concept. Frustrated, he stopped teaching for a while and instead just sat observing the way the two girls interacted. Both already used signing to communicate with each other and both, he realised, were able to convey everything they wanted to – a complete grammar of love, irritation or puzzlement all perfectly linked and all perfectly comprehended between them. And so, just as his contemporary Samuel Johnson was doing with written English back in London, the Abbé took language and began to systematise it.

Within a few years, the Abbé had begun to accept more deaf students and to establish an academy offering tuition in both signing and writing. 'Teaching the deaf is less difficult than is commonly supposed,' he wrote. 'We merely have to introduce into their minds by way of the eye what has already been introduced into our own by the ear.' He was not without his critics; others had already come up with a system called dactylology which spelt out the individual letters of each word in order to communicate. The disadvantages were clear; it was a slow method and to a deaf child, an object's written name seemed to bear no relation to the object itself. De l'Épée's signing method, by contrast, was fast, elegant, efficient and did its best to match the sign to its meaning – he could illustrate a cat, for instance, by stroking an imaginary set of whiskers, or show reverence by putting the hands together in prayer.

A couple of centuries later, there are now several different sign languages in use and fluent signers often say they prefer signing to spoken language for its ability to convey complex concepts or great depth of feeling. 'You want not speech,' as Bulwer wrote, 'who have your whole body for a tongue.'

Just as with the Abbé de l'Épée's method, British Sign Language uses signs which have an obvious link to the thing they're trying to describe: the sign for drunk, for instance, is two fingers on the right hand tottering across the palm of the left. 'Hope' is the first two fingers of both hands crossed, 'love' is both hands flat against the heart. Occasionally, the directness is a little too direct; the sign for Jew was once a hooked nose, and the sign for America had to be changed after 9/11 because it looked too much like a flying plane. The sign for Prince Charles remains a pair of flappy ears.

Amanda, one of the interpreters at the Birmingham unit, began signing aged seventeen when she heard that there was a need for more BSL interpreters and thought it sounded interesting. Her grandmother is from Sri Lanka but speaks Punjabi. Amanda doesn't, so the two of them couldn't really talk to each other. 'I thought if I became an interpreter then it would at least help someone else to be able to communicate.'

She signed up for the BSL Interpreting degree at Wolverhampton and found that it suited her – as a dyslexic, signing was so visual it felt much easier to learn. She now works for an agency who send her out to interpret all over Birmingham, which means working at the unit two days a week and then wherever and whenever she's needed – the police, GPs, the courts, colleges – 'Anywhere and everywhere. It's busy and it can be quite challenging as well.' Partly because interpreting for, say, someone Deaf who has just been arrested and who can't communicate properly with either the police or their solicitor is always going to be difficult, but also because 'Deaf people can be very blunt. It's not meant to be rude – it's just the way they say things, so I then have to culturally mediate that message.'

So what are the moments she's really remembered? 'There's memorable in different ways. Some of the worst are appointments where people are given bad news. So when I've interpreted "You've got breast cancer", or something like that – those appointments stick out. I've also interpreted for someone who had a stillbirth. That was a sad one. They'd had a baby, and the hospital were just trying to say, "Do you want us to investigate this? Do you want us to find out what caused your baby to die?" But having that conversation so soon after the birth, it was like ...' She exhales.

She hasn't done much police work but is very much aware that the way she translates has the potential to have a major impact on someone's life. 'It's having that pressure of knowing that your work affects them. So you've really got to take care with every word you interpret. So for instance with a car crash, whether you say a bang, a tap, a smash – all of those things mean very different things though the sign might be the same. And then we've got to make sure we pick the right word. We tend to go either with the person's body language, the facial expressions, the speed that they sign something or – worst-case scenario – we'll tell the police, "Look, they've used a sign which could mean this, this or this, do you want that clarified?" And then they can go in and get a bit more detail.'

IN THE STAFF ROOM of the Deaf Mental Health Service in Birmingham everyone is talking at once. The unit's two psychologists are going over the progress of each of the eleven patients on the ward, one of the two interpreters currently on duty is handing over to another, a nurse is discussing shift handovers with the admin staff and the other is talking through the timetabling of an assessment later this week.

Put ten hospital staff having seven separate conversations in one small room and it should be mayhem. But only two of those conversations are being conducted in spoken English. The other five are either in BSL or in interpreted BSL, so there's no sense of competing volume or of one voice pushing against another. Everyone seems to be conveying exactly what they need. There's a sense of motion in the room, of hands pulling shapes out of the air and meaning out of those shapes. Because BSL requires the use not just of both hands but of the face and the body, it becomes a three-dimensional language in which each speaker uses the whole of themselves.

I'm standing by the door and having an interpreted conversation with Jenny, the unit's deaf nurse. Because I can't really sign, it's taking me a while to work out how this goes. I'm asking Jenny questions but those questions are going through the interpreter Helen, so Jenny is looking at Helen rather than at me. But when Jenny replies, I naturally look at Helen because it's Helen who's speaking. So I look at Jenny all the time because she's the person I'm talking to. Jenny looks at Helen all the time because it's Helen who's turning my words into meaning, and Helen is looking between the two of us. When Helen gets to the end of one of Jenny's sentences and lets her hands empty out, waiting for the next thought to flow through her into motion, it's a peaceful moment, and it makes me want to start talking in three dimensions too.

Later that day Jenny has a moment from her duties to talk properly so the three of us find an empty room and sit down with a cup of tea. Jenny is 40 but looks about half that, bright and elegant with dark curls and an easy smile. The edge of the frames of her glasses are covered in sparkles and the fronds of one or two leafy tattoos curl out from the edge of her shirt. That sense of noticing is there again in the care she's taken to put everything together: the shoes, the well-styled trousers, the ingenuity with which she's managed to get her NHS uniform to look good.

Jenny has worked on this unit for six years, though the journey to get here hasn't been easy. She was born deaf to hearing parents who thought her lack of response to sound was due to some kind of psychological problem and who chose not to get her hearing tested. It was Jenny's grandmother who realised that there was nothing at all wrong with Jenny's intelligence; she just couldn't hear. She took Jenny to the GP, who ran some basic tests and confirmed her hunch but told her that a more comprehensive assessment would only be possible with the consent of the parents. Jenny's parents continued to refuse to give that consent, so it wasn't until the age of three that she was officially acknowledged as profoundly deaf.

Even with the diagnosis, Jenny's parents still chose not to treat her as anything other than a hearing child who was somehow making things up. Instead of learning BSL and communicating in sign, her parents both tried to teach her to speak. For deaf children, sign is intuitive because it bypasses sound to express concepts in the quickest and clearest physical form possible. But teaching speech to someone born without the biological kit necessary to process sound is notoriously difficult. A deaf child might understand instantly what you mean if you point to a cat or a lamp, but if they have no idea what sound sounds like, how can they then turn that cat or that lamp into a recognisably articulated word?

Home life became increasingly tricky. If her mother or her father needed Jenny's attention, they'd shout at her. If they wanted her to learn a word, they'd just repeat it at higher volume and then, when she struggled to articulate, get frustrated. Meeting her now, it's very clear that Jenny is a natural extrovert. But with only her grandmother providing any kind of emotional back-up during her childhood, she grew up anxious and introspective, shy of other people and wary of social contact. Jenny had no siblings because both her parents were scared that they might have another 'faulty' deaf child. They lived in a quiet area with no Deaf community and no one who she knew who was deaf. By the time she started mainstream school she had taught herself to lip-read and, because she was perceptive and clever, was speaking to a reasonable standard. In fact, school proved some kind of refuge from the situation at home. The other kids accepted her as she was and the staff did their best to make sure she got what she needed.

It wasn't until she had left school aged eighteen and

moved to Birmingham that she made the discovery that
not only were there plenty of other people like her, but that
they had a complete language of their own. When Jenny first
encountered the city's Deaf community, she also discovered
BSL. Her new Deaf friends were astonished to discover that
Jenny had managed to get so far on lip-reading and speech
alone. Over the next five years Jenny went from strength to
strength, becoming fluent in BSL, going to college, qualify-
ing as a specialist psychiatric nurse and starting work in one
of Britain's three Deaf Forensic Units – secure units for the
treatment of deaf patients, particularly those with a history
of violence. During that time she went home and told her
parents she preferred now to live as a deaf person rather than
a deaf person struggling to conform to a hearing world. Her
parents weren't pleased. Even now they still both continue to
deny that Jenny isn't just being awkward. Instead of tapping
her on the shoulder or turning the lights on and off, her
mother still shouts, and it is Jenny's young daughter who
alerts her to the fact that her grandmother is talking.

BY MIDDAY, THERE'S NO ONE in the patients' lounge except
Gregory, who's stretched out on the sofa watching telly with
subtitles. He's tall and good looking, but there's something
clunky in the shape of him, as if he doesn't fully fit the body
he's in. Gregory has been staying here for a while, and may
be here for some time to come. Having given me a sign name
– right hand down, fingers tucked in, thumb and little finger
out to the side and shaken side to side like a bell – he tells
me he used to manage a light-engineering metal components
factory with 30 staff, that he never had any issues managing
such a large company because actually he can hear most of
the time, that he's been engaged for the last 40 years, and that
he and his long-suffering girlfriend are going to get married
next year. In fact, Gregory's working life has been piecemeal,
the staff are unaware of any girlfriend – long-suffering or not

– and so far Gregory has not shown any knowledge of either engineering or other long-term employment. One of the daily tasks for everyone here is to jog Gregory back from his conviction that he can speak (his lip movements are clear, but his speech is unintelligible) into signing again.

Gregory's fictitious life may be unusual, but his light hold on BSL is not. The reaction of Jenny's parents may seem extreme but it's a relatively common experience for many people on the unit. In fact many of the psychological issues dealt with by the staff here are consequences of the entirely preventable sensory deprivation of patients when they were young. Sometimes, as in Jenny's case, it's parents refusing to sign. But, since 90 per cent of deaf children are born to hearing parents, sometimes it's just overwhelmed families with no background of deaf awareness and a lot of other children to deal with. Often, parents don't learn formal BSL but come up with their own private gestural language to communicate with their child.

Thus all the patients currently on the ward are capable of signing, though almost all learned it later in life. Toby, for instance, who's playing a video game on the patients' computer, grew up in Spain. Both his parents were deaf, but though his mother embraced Toby's deafness, his father refused to sign and continued to communicate with him through speech. Toby came to Britain aged thirteen and, like Jenny, experienced a revelatory moment when he arrived at the (now demolished) Edgbaston Royal School for the Deaf and discovered a whole world of people communicating in sign.

Maureen was born deaf but also has a late-developing condition called Usher syndrome. Over a period of time her sight tunnelled down to a diminishing field of vision until it disappeared completely. But while she still had some sight remaining, her brain tried to compensate for the lack of visual information by filling in the gaps. So, for instance, if she thought she could see the outline of a table then her brain would magically apparate a chair to sit beside it even if no chair was actually there. Because in Ushers the brain is doing its best to compensate for diminishing sight, it's a condition which often gets misdiagnosed. After all, if someone is trying to run for invisible buses or talking to Vulcans in the coffee queue then it's reasonable to suppose that what they have is not a neurological condition associated with progressive blindness, but straightforward old-fashioned bonkersness.

Which is what happened to Maureen. She was born to hearing parents who did both sign and had been living as a deaf person for most of her life until she began going blind in her thirties. When during visits to her doctor she started talking about trees in the waiting room, the GP came to the conclusion that she was becoming psychotic and so referred her to mainstream mental health services. It was only because she was also deaf that she was then referred to this unit, where someone spotted the discrepancy between true psychosis and Ushers and began giving her the treatment she needed.

Maureen is sunny and cheerful, happy to chat for hours and keen to talk about her upbringing in a rural area with no Deaf community, and her daughter, who is hearing but bilingual in English and BSL. When I ask Maureen a question, Helen the interpreter conveys it by taking her hands and allowing her to physically trace the shapes that those words make. There's something gentle and intimate about that taking of hands. All through the day I keep being reminded of how very un-British deafness can be, with its open, easy physicality and the absolute obviousness of touch. Speech, by comparison, begins to feel stifled, as if all hearing people are doing is focusing on the things they're trying not to transmit.

ANDY HEARN is a senior software engineer who I first contacted because of a blog he'd written about his commute to work. His deafness is of a type known as Connexin 26, a gene mutation which causes deafness with no other symptoms. He's the fourth Deaf generation of his family, though both of his sons are hearing.

The first time we met was in a crowded pub opposite Victoria station – travellers loaded with suitcases, office workers going for a pint, theatregoers in sparkly dresses meeting their dates. Andy had come straight from work. He's late thirties with a long face, a trim beard and a hearing aid behind one ear. His eyes have laughter in them, and an open intelligence. He was sitting at a table facing outwards towards the door, perfectly positioned to see what was going on but directly below a speaker bellowing out music. Over to the left was a TV screen relaying BBC News 24 muted with the subtitles on, and a couple of games machines were beeping by the door.

Neither my lip-reading nor my signing is good enough for fluent conversation so I got out my laptop and we sat next to each other reading what we'd written as it came out, with all the laughs and reactions and 'No! Reallys!?' included. Sometimes either he or I would sit there for a second, fingers hovering over the keyboard, trying to work out how to express an idea or a thought, and sometimes both of us were clearly itching to grab the laptop and respond to something the other had said. Some parts were serious, though much of it degenerated into 'Everything You Ever Wanted to Ask a Deaf Person But Couldn't Figure Out How To Ask'. Real-time conversation in print was an entirely familiar concept to Andy, but I found it exhilarating – a dialogue full of gestures and laughter conducted silently in the centre of a mess of sound.

When I'd been talking to Amanda, the BSL interpreter, she'd said that the police occasionally get called to pubs or bars because people find signing too 'aggressive'. Is that really true?

'Yeah,' writes Andy. 'True. Police have been called to disperse us Deafies. I've been involved a number of times. Looking back, and despite the fact that some of the pubs discriminated against us, we were hard to evict after closing time. We descended on pubs in hundreds – sometimes thousands for the bigger places.' The problems generally started at closing time. 'We hang about, we're slow to leave the premises, the police then arrive and order everyone out. We've had something to drink, so we think it's a good idea to try getting the police to sign.'

What follows can then become a splendid example of double-sided miscommunication. Faced with a large group of people waving their arms around and apparently refusing to respond to direct orders, the police wade in. Because they're trained to deal with gang situations in which direct eye contact is seen as confrontational, when they think a situation is about to escalate, they look away. 'But that only works us up. An incident at Waterloo stands out. We had to leave the pub and one guy wanted to finish his drink, so he walked out with a glass in his hand. The policeman by the door told him that he couldn't do that, but obviously the guy never heard him, so the police tried to arrest him. Us Deafies, being the nosy lot we are, started to move forward to see what was happening, so we were all closing in, the police were screaming into their two-ways, "Back up now!", the vans and dogs arrived and then a few of the dogs started becoming friendly with some of the Deafies who had been on animal welfare courses – we suss out animals better perhaps.' He grins. 'It all got a bit out of control.'

Another glass of wine, and we go back to the reason I'd first contacted him. Hearn's daily route to work takes him through Gatwick Airport and over a period of several months he kept finding himself being pulled aside to be searched by airport police. The first thing he'd know would be a hand on his chest and an armed officer moving him out of the stream of people up against a wall. Once he's in a stress position – and now with a large group of spectators surrounding him – he'll be searched. Though he tells them repeatedly both in sign and speech that he's deaf, the information appears to make no difference until the police have found his wallet and driving licence and phoned back to base. After that, they become a lot more friendly. 'I suspect that they realise that I'm not who they want, or I don't have a criminal record. At the end they just give me a piece of signed paper to prove that I've been searched in case another squad stops me. By that time I'm just relieved to be let go.'

He's been pulled over when driving too. How many times has this happened?

'At Gatwick, I think five or six times. Driving – three or four times. Security portals – loads of times. It was almost guaranteed until I became a father. Having kids with you is a great way of slipping through. I think now each time I expect it, so maybe my behaviour changes slightly.'

But what is it that the police see in him? Hearn says he looks serious and walks quickly, but then so do a lot of people in airports. What may set him apart is his attention to human and spatial detail. Hearing individuals are often stressed by the whole business of airports and flying, but that stress generally reflects inwards. If you're deaf, you're focused outwards, registering visually all the information which might otherwise come to you aurally – announcements, warnings, communication from those you're with. Those who can hear are accustomed to using their ears as eyes in the back of their heads, so if they're walking through an airport and a luggage cart comes beeping up behind them then they move out of the way without needing to turn around to verify the cart's existence. Hearing has given them a three-dimensional comprehending. But those who are deaf can only be aware of that cart because they've seen it. One sense must do the work of two.

So what was it he did to change his behaviour? 'Now each time I see an armed cop, I take a deep breath, try to relax, and walk looking straight ahead.' In other words he no longer does what would be natural to him. Instead, he behaves like someone hearing – he keeps his lights dipped and his eyes on the destination. 'They've left me alone, though I've caught a double-glance at me now and again.'

Deafness is as much a part of Andy as maleness is, so it's almost impossible for him to isolate the ways he understands the world differently. But, he says, 'Our other senses do become enhanced when one goes. I think it was last week when [he and his wife, also deaf] were driving, we were in slow traffic and I was the first to move out of the way for an ambulance. I reckon we catch the flashing lights before others do. And I have to be very careful where I look in very crowded places such as the Underground – we Deafies glance slightly longer than the next person and are more sensitive to the periphery of our vision. At work I used to sit opposite a guy with Tourette's. The gesture kind, though I've caught him coughing over his swearing. Anyway, each time he jumped up to wave or fist the air, I kept on being distracted, yet colleagues next to me weren't bothered. Comes with growing up with people waving to get attention I suppose.'

So, I write, can you explain something to me? How do you communicate in situations where you've got your arms full? How do you sign while driving, or when you're holding one of the kids? Andy grins. 'I've had hearie passengers pointing to the road ahead when we Deaf are in the front. But with signing in cars, again, the peripheral vision comes into effect. Apparently, we're statistically safer.' As for domestic duties, 'There's an American comic strip drawn by a deaf dad and written by a hearing mother. One strip had the guy asked by his mates how come his wife let him go to the pub so often. The reply was, "Ask her when she's doing the washing up."'

Andy has been MRI-scanned a few times as scientists search his brain – his fourth-generation brain, now totally adapted and enhanced – to work out in what ways his sensory topography has been remapped. 'Anything for languages and science, me,' he types. That extrasensory perception also has its uses within coding. 'My work colleagues [all hearing] do think more sequentially. I have to craft my emails more carefully for them. My whiteboard scribbling is more pictoral than organised and mind maps suit me more. I'm not sure whether it is me or it is because of my spatial thinking, but others seem to give me the most devilish software bugs to hunt down. But when I've caught them, the reward is … amazing.'

So what was it like when your first son was born, and you realised he was hearing?

'He was born six weeks premature so we were worrying about his health, which meant the state of his ears was put aside. By the time he got strong enough to leave the SCBU, we knew he could hear. Ironically his hearing is "exceptional", ten decibels above average.' Both he and his wife felt a certain ambivalence about the break in the genetic line. His sons are bilingual in spoken English and BSL, so 'they appreciate the signing and being able to code-switch', though he knows that if they then have children who are hearing, it's probably going to be difficult to maintain that culture beyond one generation.

He and his wife, on the other hand, were brought up with an unbreakable sense of belonging. 'The best way to describe it, I think, is that when we [he and his wife] were globetrotting in 2006, whenever we came across a Deaf community, we "belonged", no matter what the written language, and no matter how far away. Nepal, even Lhasa. We went through Russia and China/Mongolia/Tibet where the script isn't Latin-based just fine, and had a brilliant time watching our other hearie travellers trying to gesticulate. Deafies from other countries have their own sign languages, syntax, grammar, et cetera. But because it's visual and lends itself to iconity easily, we can quickly grasp the more abstract signs. When deafies from different countries meet, they start to use "international sign"' – not an official language, but something universally understood. 'It transcends all creeds and all other distinctions. That too is a recent revelation, something I took for granted until fifteen or so years ago. Deaf first, nationality second.'

The third time we met, Andy Hearn suggested the Churchill Arms on London's Kensington Church Street, a pub so festooned in flower baskets and frilly-knicker flags-n-jackery it felt like drinking inside a presentation bouquet. They didn't have a table, so we walked down the road to a restaurant that served good steaks. The previous couple of times, Andy had picked places which were handy geographical meeting points – cheery, good for swinging the laptop between us, close to the station or the tube – but which were difficult audiologically – shouty pubs or very busy restaurants.

I know it seems a bit slow, but I don't think I understood until that last meeting that Hearn simply doesn't assess things on the same basis that hearing people do. If you're deaf, then all sound, of whatever volume, is irrelevant. It's not there. It's just not a factor. But if you're deafened, then that snarkish hunt for the right sort of sound can often occupy a lot of space in your life. So while even now I reflexively search for a good acoustic environment (quiet café, low-ceilinged room), it's a consideration that doesn't even feature for Andy – all he'd need to figure out is whether he likes the place.

It took a long time for the penny to drop, but I got there in the end. The difference between deaf and deafened is that, for the deaf, there is no lack, no loss. There's always just the world as it is, complete.

11

Vision

BY 2004 I HAD A PART-TIME JOB working in a pro photographic lab off the Kingsland Road. My regular commute took me from west to east along the Marylebone Road towards Old Street. Depending on the weather I'd sometimes get the tube, sometimes the bus, and sometimes cycle.

Marylebone Road was originally designed as the northern side of London's first bypass, though the tarmac soon flowed over it, moving onwards and absorbing the road into the city's great blank mass. From Edgware Road in the west along its Euston Road continuation to Pentonville in the east, there are along its length five hospitals, one national library, a fire station, four major rail termini, seven tube stops, the Methodists, the Quakers, a school for the study of astrology, several churches, many doctors, one overpass, three underpasses, a lot of plane trees, several of the best musicians in Britain, a booth for measuring the city's daily pollution levels, a courthouse, an apple tree, a waxworks museum, a famous consulting detective, an underground river, several razor-thin slices of human brain and a necklace made of teeth. There are a few pubs, hotels and shops at either end of the road, but – perhaps because the road is also a three-lane highway – it's not a place to linger. But its sheer abundance, its utter essence-of-London-ness, make it a good spot to pick out the city's soundtrack.

In BSL the sign for London is the same as the sign for

noisy, though by the round-the-clock standards of other great international cities – Mumbai, Buenos Aires, New York – London has always had a muffled quality. Thus the Marylebone Road is about as loud as this place gets. Because of all the hospitals and fire stations along its length there is almost always the sound of a siren rushing westward or eastward, and each of those sirens has a song. At the western end prison vans scurry down Lisson Grove heading for the back of Westminster Magistrates' Court, giving no more than a quick woop-woop before vanishing behind high walls. Military police bang through the traffic on their way to a tea break while fire trucks cry havoc across the city and fire engines bawl northward.

And then there are the ambulances flashing their way to good news or bad, up from St Mary's or down from UCH. Pre-deafness I'd be cycling back from work along the bus lane and suddenly there would be a sound loud enough to spring me six feet into the air. Sirens don't just have to rise above the ambient noise of a city, they have to reach the ears of a driver in the sealed cabin of a far-off car. Because drivers can't hear outside sounds well, the volume of those sirens has been increasing until they're now often set at around 110 to 120 dBs. Cyclists and pedestrians aren't in sealed cabs and can obviously hear fine, so the noise of the ambulance sometimes creates behind it a bow wave of figures leaning away or covering their ears.

Other sounds, some of them seasonal. The shush of rain under tyres over the roar of leaf-blowers. The ticking of hot car metal as it expands under the sun. The unsynchronised click of stilettos. A beggar outside Lloyds bank, propping himself up against the wall: 'Yous! Gies yer change, ya fuck!' The gusty sigh of a bus's air brakes against the shaking of cabs at the lights. The collective rush of trucks passing. Stop. Start. Stop again. Fragments as people pass: 'Most of salami is, like, donkey. Or sometimes horse.' Pigeons, their wings beating against the leaves of the trees as helium balloons sag between the branches. Two armed Met policemen, all stab vests and jackboots, discussing bargains on the QVC shopping channel. The shudder of the Metropolitan Line beneath my feet. Tourists queuing outside Madame Tussauds, calling back their children. Workmen putting up scaffolding; the flat notes of boards and poles being set in place, the chink of the couplers thrown down on the pavement. The rattle of fighting magpies. 'Yeah, well, he talks about it so much, it's obvious he's a virgin.' Musicians wrangling coffin-sized double basses over the road, yanking at their trolleys before the lights change. A Lamborghini gunning in frustration. Somewhere behind, the diesely snort of the trains at each station. And far beyond that, the sound of London breathing.

Sitting on the top deck of the 205, I would listen to the schoolkids, their voices raised to grandstand.

'I was like, Un. *Fuckin.* Believable. I mean.'

'Yeah, an really I done my mum a Hollywood the other day.'

'Likefuckinwellsickyeah?'

'Not like I mean what that's wrong, but I'm saying by now you should be pissing orange.'

'I mean, whodefuck?'

'He was like, yalright? I was like, yalright?'

'Are. You. Fuckinshittinme?! Omi*god*, Shan, youfuckinnevah!!'

Being both professionally and personally nosy, I could never get enough of it. In the evening, staring out of the window, I'd gobble up romances, cliffhangers, showdowns, conversations about what to cook for dinner that night, one-sided phone conversations, arguments. *How old's the lasagne? Well, let's use that before it goes off.* Replays of stand-offs with a landlord or parent–teacher assessments. *The wiring – omigod you should see the wiring in that place. It's a joke.* A war on YouTube. Music leaking from someone's headphones. Unresolvable health crises, battles with doctors or fights over medication. *This new one, I don't think she even read the bloody notes. I mean, now she got me on twenty mil of amitriptyline and if I've told them once I've told them a thousand times, it don't work for me, that stuff.* The politics of offices. *D'you think we're talking a total rebrand?*

Sometimes the voice behind me would be flat and monotonal, drained of colour or inflection. Sometimes it was animated or coy. Sometimes, from the short sing-song – *Hello, pet! Are you having your tea?* – they were talking to a child. Sometimes it was intimate, face turned away to the window, phone pressed hard against their head. It didn't matter. I didn't care – I'd just inhale all of it, whole box-set dramas at a go. Chats in languages or dialects I couldn't even recognise, each from a different region of the body. French came from farther back in the throat, the pitch of German was deep and steady, and Russian was deepest of all, dragged up like bass from the depths of the diaphragm. Listening to an older man talking Russian, I could practically see the sound waves rolling out of his throat. For all I knew, he was discussing crisp flavours, but if he was, then those flavours sounded as rich and terrible as Montagues and Capulets. The difference between Spanish and Italian was the difference between an Edinburgh accent and a Glaswegian one – one pushed forward, higher, right to the top near the nose, the other swinging off the tongue, instantly recognisable. It was a game; could I recognise the subject under discussion even if I couldn't understand a word of what they were saying? Could I at least get the region if I couldn't get the sense? And, if I could get both the topic and the language, then could I work out what was likely to happen next?

THAT WAS BEFORE. And then there was after. I could still cycle or get the bus from west to east, but though I was taking exactly the same route, it had changed completely.

Imagine London turned down. Cut out the traffic. Cut
the trees and the pigeons. Cut the leaf-blower, the trains, the
smoothing rain. Cut the air brakes, the scaffolders, the click
of heels. Cut the beep of a reversing truck or the bang of its
shuttered back. Cut the air-brake exhalation of the bus. Cut
the kids outside Madame Tussauds or the chat of passers-
by. Cut the angle-grinder's rushed complaint and the rise of
a motorbike's frustration. Cut the tourists. Cut music. Cut
conversation. Cut Korean, Scottish, Arabic, Spanish, English,
American, French, Estonian. Cut the occasional shout over
the traffic or the bark of a dog. Cut the shriek of a black cab's
brakes. Cut the whole lot. Cut everything except the sirens
and the wails of unchanged babies. Silence it all.

Or rather, take it all down by about 80 per cent. Take out
all of the juice and most of the pith. Remove half the sense
and flatten the rest. Leave what remains as a disconnected
sequence of hisses and sibilants. The edges of sound are still
there but the sense in its centre has gone. I can still feel the
vibration of the bus and the windows shaking slightly as it
stands. I can see the drills and the grinders but the sound is
stopped off. I can still hear the sirens but until they get very
close I can't separate them from the mutter of the engine.
I can't hear London breathing any more. Perhaps it's dead.

None of this is unpleasant or uncomfortable and,
because it's happened over a long time, I've adapted to it.
But it is strange. I know some sounds because I catch the
end of them, like catching the last words of a well-known
quote or phrase. I can hear the edge of a diesel's idling motor
not because I can really hear it, but because my brain knows
the sound of it so well it completes the missing phrases. I
know the station announcement is a station announcement
because of the rhythm and the distance, not because of the
words. But other sounds, disconnected from their source
and from any surrounding links in the auditory chain, make
less sense. They just appear as pitches or tones lost in their

own space. Half a word in a sentence. A slice of ringtone. A shout. A sudden metallic bang. The tail end of a longer sound like the squeaky shriek of wet bike brakes or the thump and rattle of plane tree branches against the roof of the bus.

The closest analogy would be to imagine putting on ear defenders and then listening to a radio on the other side of the room. You can't hear much, but you can hear the difference between the fenced-in sentences of a news bulletin and the ramblings of chat. If it's a music station, you can pick up the bass line but not the melody. If it's sport, you know exactly where the goals are in a match report or a horse race rising to the finishing post. Instead of just receiving the sound, as you would do if you could hear fully, it's now an entirely interactive experience; you're given a certain level of information, you work out the rest from that. Is the news politics, foreign, or crime? Can you recognise individual presenters or find the shape of a whole song from one single strand? Can you, equipped without hearing but with all your experience and knowledge, colour in the spaces left by sound?

When it got to the stand at Euston the bus would loiter for a while beneath the concrete canopy. I'd sit with all the other passengers on the top deck looking out over the big 1960s forecourt while the drivers stood and chatted. Outside on the forecourt were the usual mixture of travellers, junkies, dog-walkers, tourists, skater kids and students. I'd look at all these different tribes and start to read. Those two, the couple who have been sitting near the top of the steps. They look like they're in love, but it's always him who moves towards her. She sits a little tipped away from him and she's not touching him, she's wrapped over her phone. The man there with his bag on the ground, talking to the older guy. What's their relationship to each other? Is there some kind of deal going on? From the bag man's big punctuating gestures it looks like he's trying to persuade the other one of something. The retired couple standing slightly apart beside their unscuffed suitcases. Where are they going to? Have they argued? He moves stiffly, like he can't turn his head, like there's always something he doesn't want to see just to one side. Two male paramedics having a coffee, one of them bending forward in the chair as if his back hurts. Have they just come off a shift? Do they like each other? Look at the queue for the 68. There's nine people there. How come some of them really stand out, and some seem always to be part of the concrete? That smooth-haired businesswoman, walking across the forecourt towards the road with a tissue in her hand, never looking up, turning her head from the other passers-by.

A man comes out of the station, three young boys trailing behind him. His shoulders are festooned with kids' rucksacks and he's trying to manoeuvre a larger case with a broken wheel. Halfway along the forecourt one of the boys drags harder and harder on his hand until he lets go and sits down on the pavement. Then he lies down. The man tries to pull him up again but the boy isn't having it. The man stops and one by one takes off all the bags he's carrying. The other two boys have vanished out of sight. A lengthy period of arbitration takes place, intense diplomatic efforts focused on the effective conveyance of mutually beneficial long-term goals. A bus – the bus they're all supposed to catch – arrives. The man points at the bus. The boy on the ground gazes up at the sky and his face turns red. It is March, and chilly, but it is only when demilitarisation talks reach a chicken clause (man gnawing imaginary drumstick) that the boy sits up. By the time the man has successfully concluded negotiations, gathered the other two and run towards the stand, the bus has moved off.

Over on Eversholt Street, a courtesy bus carrying tourists from one of the hotels has shunted into the back of a minicab. Both the cab and the van have stopped, blocking the road, and the courtesy-bus driver is walking over to the cab. Both he and the minicab driver are giving it their best, shouting and finger-stabbing. Their faces are bunched up and their movements are sharp and hard, their meaning clear. Behind them the driver of the bus is leaning out of his window and gesticulating at them to unblock the road. The minicab driver is accompanying all his insults with a peck of his forefinger and a jerk of his head. The courtesy-bus driver has his arms open wide and is sketching out through the air the absolute and spectacular enormity of the minicab's stupidity.

I would watch all of this, willing the bus to get moving but entranced by the morning's impromptu theatre. And the longer I watched it, the more something became evident. I was seeing all these people, their conversations, their non-conversations, their intentions and anxieties, and, four times out of five, I reckoned I could make a reasonable guess at what was going on. Who knows? Unless they got on the bus and sat right next to me I wasn't going to be able to fill in the factual gaps. But if I looked through the silence – really looked, with the whole of my attention – then I could see the hubbub of interaction. I could see desire or stress or jealousy or frustration, I could see how many people there had already left before they'd gone. I could see the way homelessness rendered men invisible or the sheer, ceaseless 24-hour hard work the addicts put in to feeding their needs. I could see the gap between who people wanted to be, and where they really were. The odd thing was, I couldn't hear a thing, but I was having no difficulty in understanding every word.

SOME TIME IN 2004 there was an exhibition at the National Portrait Gallery of photographs by the painter Lucian Freud's assistant and model David Dawson. Over the years the two had worked together Dawson had gradually amassed a whole series of photographs of Freud at work. The images were casual and affectionate, almost always taken using natural light, some in his studio but others at private views or in other parts of Freud's house in London.

One image showed Freud's friend and fellow artist David Hockney sitting beside the easel on which stood Freud's portrait of him. He looks like he's just thinking about cracking a joke. Beside him is an ashtray and behind is a pile of dustsheets and a thick swarm of colour where Freud has dabbed at the wall to test the texture of his paint. Freud himself is just walking through the door, white shirt on, scarf knotted around his neck and paintbrushes in one hand. He's looking into the camera, and the look he gives it drills straight through the lens.

A few images on there's a picture of Freud standing in a stable beside a grey horse. One hand is on the horse's halter, the other tucked behind its mouth. The horse and the man stand beside each other like friends, the stable dark behind them. Freud's gaze is as straight as a sparrowhawk's. I looked at the two pictures in the exhibition, and I thought, 'That man is deaf.'

It was an odd thing to think. Though David Hockney does indeed have severe hearing loss, as far as I know, Freud never went deaf. But what seemed to be visible in the photograph was the degree to which Freud had sharpened one single sense to such a level that it became the point at which all his energy and most of his being was focused. Everything, to Freud, was there in his seeing. He didn't need hearing to be a great painter because hearing couldn't help him capture the soul of a sitter. All he wanted – all he needed – was a ravening appetite for light.

Hockney, who's now been painting for almost as long as Freud had, has said as much. 'I actually think the deafness makes you see clearer,' he said in a 2003 interview. 'If you can't hear, you somehow see.' It seemed to me that Freud's sight so absolutely dominated his other senses that his hearing had receded.

Many years later, I realised that what I'd seen in those photographs was the same thing I saw in the faces of the deaf and which Andy Hearn had described. It was the difference between looking and seeing. Their eyes were switched on, their gaze alive, and there was an acuity in their observing which seemed striking partly because it was so rare. If every one of us has a light at the back of our eyes then it often seems as if most of us choose to live with that light dimmed down. Real seeing is a raptorish faculty, a sharp, hungry tool, strip-mining information from the visual world. Freud had it. The deaf have it. Children have it until they get taught to turn it off. We all have it, but that open seeing is also such an intimate gift that it seems too personal for daily use. *Don't stare.* Why not? There's so much to be curious about, such sensory pleasure to be had. Freud looked like the deaf because only the deaf choose to see.

Looking at those photographs, I realised something. I might be losing my hearing, but I still had everything I needed. Sight gives you the world, hearing gives you other people. Take away hearing and what have you got? You've got the world.

PRO PHOTOGRAPHIC LABS ARE – or were, at least – fantastic places, full of skill and gossip. I started work there just at the point of maximum disruption as film gave way to digital, though this company was set up for both: two separate spaces for colour and black-and-white film printing plus a further space within the same building for digital output.

Though the volume of film coming through was declining by the week, the lab's four printers still had plenty of work and each remained the undisputed overlord of his or her own darkroom. One by one they'd pop like bubbles through the black revolving door, re-emerging a few moments later to lean against the huge colour printer, marking up negs and complaining that the milk was off. Dave had once been an architectural modeller and, when someone bought a small plastic model of Gollum, he only had to add a faint yellow tinge to the eyeballs and a few thickened veins to produce the spit of himself. John was Irish, a big softie who produced his prints like magic tricks from a state of chaos, and over in the corner was easy-tempered Mike, a Stones fan who ran the processing side of things.

Back at home I set up my own little black-and-white darkroom in the bathroom and taught myself the basics of monochrome printing. I'd get back from the lab, take the dog out and go straight in: *Just a couple. Just the portraits. Just five minutes.* Time became something measured out in elephants (*oneelephant, twoelephant, threelephant* ...) or passed unmarked in three-hour gulps. I'd come reeling out at 2 a.m. with a headache like a sledgehammer and then be up at seven to pull the prints off the washing line and then get in to the lab to do the whole thing all over again. I was stoned on photographic chemistry, pickled in dev, dissolved in fix. What I wanted, again and again, was that Excalibur moment when an image – a road, a face, a flame – rose up from the water to meet me.

It's easy enough to learn black-and-white printing at home but colour printing is almost impossible anywhere except a lab. Instead, I did my best to pick up as much as I could of the old lore from the printers themselves, men (usually men) who had spent their lives mastering an art which was now expiring before their eyes. It wasn't just that photographers and editors were all swapping to digital, it was that the big photographic manufacturers – Kodak, Fuji, Agfa – were reducing their stock of film and chemicals to the point where a whole delectable menu had become a shrivelled little snack.

Still, from my point of view, it was a privilege just to watch them. Part of the art of colour printing is in having two very highly developed senses: sight, and touch. Printers have to be very well trained at looking but they also have to be good at working just by feel because once the light-sensitive paper is out of its packet the darkroom has to be totally dark. Neg out of its sleeve into the enlarger, find the right paper box, pull the paper out of the thick plastic sleeve, locate it in the mount, adjust the dials, notch more magenta here, degree less cyan there, couple of exposures, out the door and into the machine – all of it done by fingertip.

As a visual education it couldn't have been bettered. In amongst the boxes of corporate portraiture or catalogue fodder, the lab processed work by some of the best photographers of the day – Juergen Teller, Tim Walker, Sam Taylor-Wood, Josh Olins. The sheets passed through our hands tagged with names, dates and job descriptions, passing along the lightbox to the next stage in the process. There was a rhythm to it, a smooth, dust-free motion from uncurling the negs to slipping the last line into the sleeve. Once in a while we'd reach for the lupe, bending over a single frame to check for errors. Sometimes film came in bearing the telltale fog of airport security X-rays or the distortion of dated film stock. Some had clip-marks or rips or the single shaft of false sun where the lens had slipped and the light had got in. Sometimes the photographers themselves had a single defining trick – always taking on the diagonal, perhaps, or preferring underexposure.

And some – some you could just never stop looking at. Mostly, those were the ones taken by Juergen Teller. There he was, the man himself, line after line of naked Tellers standing there with nothing on but a bratwurst. There was Charlotte Rampling in the Hotel Crillon with Teller upside down on a grand piano beside her. There was a blank wall with a sapling growing out of the cracks. There was Kate Moss stuffed into a wheelbarrow or Victoria Beckham's legs spilling out of a Marc Jacobs bag. There were a hundred flaming Vivienne Westwoods laid out like invitations. There was snow, melting, or a fragment of ragwort on a rainy pavement. Teller again, caught in filmic whiteout. His images were irresistible partly because everybody was naked and doing strange things with food, but also because they were great – funny, provocative, brilliant, perceptive, revolting. They told you about what both people and film could do when they were pushed.

And then over in the other corner someone would hold up Tim Walker's looking-glass images of galleons and fairy-tale girls, correcting the depth of a shadow or pointing out nuances that I hadn't even considered. All day, as I filed or cut or sleeved, I gazed at this rolling procession of images, good, bad, dull or brilliant. If I wanted more detail I could ask the printers how it was done or compare a dark print with a light one. Pulled halfway between Mike's bar-by-bar critique of *Goat's Head Soup* and the radio's foretelling of calamities at the Catthorpe Interchange, I'd barely register the arrival of the photographers and their entourages. The printers would clamp each new print to the wall and the waiting crowds would circle. 'Top right,' someone would murmur. 'Blown.' A collective pause. A squint. The waiting eyes flicking between it and the print beside it. Too contrasty. Too flat. 'Maj is off.' Maybe try it with a keyline? Or gloss? Or 16x20? The printers would flip back behind the door, waiting there in the dark by the machines until the crowd wandered off.

Because of the background hum of the machines I'd have to work hard in the lab to hear. But it wasn't sound I was concentrating on any more, it was vision. In there I could get hooked on looking, I could stare to my heart's content, I could teach myself a whole new communication. Cycling north in the evenings, deaf to the traffic swirling around me, I'd think, I am happy. Somewhere in all those pictures of legs and handbags, I had found my way out.

12

Surfacing

ONE DAY IN THE SUMMER of 2009, I went for my regular appointment with the audiologist Jacqui Sheldrake. But this time, instead of just taking the standard audiogram and adjusting the amplification on my hearing aids, she ran a different series of tests. This wasn't just the usual headset-on, clicker-in-hand pure-tone audiometry, this was bangs and tuning forks and things against the side of my head.

When she'd finished, Jacqui sat back. Behind her was a fireplace with a dark Edwardian mantelpiece and emerald tiling round the edge. Once, the mantelpiece had displayed only a few items, but over the years the silver-framed family photographs had been joined by a small tribal grouping of international travel ornaments, pencil sharpeners and defunct phone chargers.

'You should go to France,' she said.

'*France?*' I said.

'I think,' she said, picking her way across each word with even more care than usual, 'That what we're looking at here is otosclerosis.'

I shook my head. Whatever it was, I'd never heard of it.

'Remember there are two kinds of hearing loss? There's sensorineural, which is the one most people have where the cilia in the cochlea wear out, and then there's conductive?'

I nodded.

'Sensorineural loss takes place in the inner ear. If you

had sensorineural loss the probability is that your high frequencies would be poor but your low frequencies would be better and you'd struggle to hear in very noisy places. But you don't, do you? Your low frequencies are poor but your high frequencies are better and you've always said you can hear fine in very noisy places.'

She pulled her keyboard towards her, tapped, and angled the screen around. On it was a diagram of the middle ear, showing the malleus, the incus and the stapes. 'Otosclerosis is a condition of the middle ear, not the inner. What happens with this is that bone starts to build up around the stapes. And as it builds up, the stapes can't move properly which means it's not vibrating against the tympanum and therefore it's not conducting sound properly. With me?'

'Mmmm.'

'The tests I just did were for conduction, not for sensorineural loss, which is what we've been concentrating on previously. If otosclerosis is not that severe then it can be confused with something else, but this time it seems to be coming up loud and clear.'

She paused. 'With otosclerosis there's a very specific window of opportunity where it's severe enough to be easily identified but not so severe that the hearing has completely gone. I would say you're down to about twenty per cent of normal hearing now, which is still within that window.'

'Oh,' I said.

'In France, there's a group of people who have been doing private operations called stapedectomies on people with otosclerosis for a long time. There are people here in the UK who do the same operation but they don't do it often, and most of them will have been trained by this French lot. They pioneered it and they're really good at it because that's what they do, all day every day.'

I felt heat sweep through me. 'Could this improve my hearing?'

Jacqui made calming motions. 'The results they get are mixed. Some people find the operation works amazingly, some it makes a difference to one ear or the other, and I should tell you that for some it makes it worse. But why don't you do a bit of research and see what you think? I'll write a referral letter and we'll go from there.'

I was hot and cold, neither here nor there. I cycled home, thinking about what she'd said. *Don't be daft – there's no such thing as an operable ear condition.* I didn't know what I knew any more, and I didn't know what I was. Dithered. Walked upstairs. Walked downstairs again. Made calls, spoke to my partner Simon, wrote emails, tried to put it out of my mind. Sat there at my desk with a pad of paper in front of me drawing the same thing over and over again. *Maybe there's a cure. Probably there is none.* Googled otosclerosis. *Otosclerosis: A hereditary condition causing progressive deafness due to overgrowth of bone.* Took the dog out. Came back, googled it again: *medial fixation ... stapes ankylosed ... multifocal areas of sclerosis within the endochondral temporal bone ...* I didn't understand, and whatever I did understand I certainly didn't understand enough. I called my sisters, talked it through with them. This wasn't cochlea implants, this was something else entirely – some kind of internal prosthesis thing; I didn't know. Oh, right, said Flora, that sounds great. So when exactly do you get these cognitive implants?

A couple of days later, I called a patient of Jacqui's who had already had the operation. Adam Sieff is a record company executive who started to develop otosclerosis in his thirties. When Jacqui had told him about this clinic in France, he hadn't taken long to make the decision.

'If you go for it,' he said, 'then they don't do both ears at once. They operate on the worse ear first, and they like you to leave at least six months in between.' He and his wife had travelled to France for the first operation two winters ago, and he'd had the second operation the following summer.

The surgeons operated the day after he arrived, he'd spent nine days recovering, and then on the tenth day, they'd taken the bandages off his ear to reveal his sparkly new hearing.

'And did it work?' I asked.

'Yes,' he said. 'Absolutely. Not perfectly, and I'm still using hearing aids, but whereas before I was down to about thirty per cent of normal hearing, now it's more like sixty to seventy per cent, and stable.'

'What kind of difference does that make to your life?'

'On a practical level it's just made everything that much easier. And on an emotional level, it's huge. I'm still adjusting to it.'

'So do you think the operations were worth it?'

'Yes,' he said. Unequivocal. 'It's not perfect, but it's a damn sight better than what I had before.'

I put the phone down. My head hurt. My head seemed to hurt a lot these days. Even that – just the making of a phone call – was getting harder. It wasn't just that the volume never seemed to be high enough, it was the murk and drift at the end of the line. Unless it was someone I knew so well I could second-guess their diction or use my image of them to fill in the visual blanks, I'd got so used to reading someone's expression on their face that it was difficult to do without it. With someone like Adam whose speech pattern I didn't know, I'd end up with the receiver rammed so hard against my ear I felt like I was trying to pour myself down the wires.

There was an operation. *There is an operation.* Oh my God!! *There's a bloody operation!!*

After work that day I met Simon and as we walked across the park in the dusty summer sun, I told him about the conversation with Adam.

'It's expensive,' I said. 'Around seven thousand euros.'

'What do you want?' he asked.

I stopped on the path and looked at the dog, who was leaping through the golden grass like a shampoo advert straight into the waiting jaws of a Doberman. 'This isn't small stuff,' I said. 'This is head surgery. Twice.'

'Do you want things to stay the same?' he asked.

'No.'

'So go. Go and see.'

IN A ROOM in a town crossed with trees and pharmacies in southern France, an old man pushed a little box across his desk towards me. Inside, lying like a jewel on a special moulded lining, was a tiny titanium hook on a stick. It looked like a long metal question mark. I picked up the box, bringing it in close. A question mark, I thought; how apt. The stick was tiny, perhaps five millimetres long, and for a second I felt only stillness. I saw myself in X-ray, the flesh of me vanishing into translucency, nothing visible except my bones and these little crescents. Like earrings, but inside.

Four months later Simon and I set off on the Eurostar, heading down France towards Montpellier. Trees, ploughed fields, frost stilling the grass. Crows in the fields and the ploughed distances of the north. Telegraph lines, looping past to a rhythm, *swoop, swoop, woop, woop* ... I sat by the window and looked down at the pattern of hairs on Simon's arm, overwhelmed by his aliveness. In 48 hours, I would be lying on a table while someone cut holes into my head. What was I doing, risking the sound I'd still got?

There are two records of the next fortnight. The first is the chronological one – timetables, check-ins, surgery, reveal – but the second is always there behind it, an alter-image burned out of extremity. The crimple of waterproof hospital mattresses and the gangrenous stink of microwaved stews under the plastic cloche. The slop of slippers on lino. Low north winter light flickering through the plane trees along the Canal du Midi. Simon, arriving and leaving from his hotel near by and the expression on his face as he looked at me, lop eared, cross eyed. Walking slowly along lines of

winter vines, their branches stumped, looking at the desic-
cated white soil beneath my feet.

Dr Vincent's arrival on the ward was always marked by
a shift in air pressure. A neat man in his late fifties, he wore
the same crocs and comfy white fleeces above his scrubs as
the other medical staff, but when he emerged from his office
and walked down the corridors, something seemed to swirl
in his wake. While the rest of the clinic was rendered in standard plastics, his own office was full of light and plants, a
civilised space away from the impersonality beyond. He had
been doing these operations for years, he assured us. Sometimes the results were good, sometimes they were very good,
but there was almost always a significant improvement to
hearing. The risk of side effects was very low – only 0.5 per
cent – and he was proud to say he'd never yet had a patient
who had regretted the operation. The only tricky thing for
foreign patients was the trip back home. To get to the stapes
he would cut a flap in the skin beside my eardrum, which
meant that the cut would need time to heal. To give it a
chance to do so successfully, there must be no alteration in
air pressure around it for the next month or so. Which in
turn meant that I couldn't fly or take the TGV home, and
that I must stick as close to sea level as possible.

On the second day he operated on my left ear. When I
woke up I was told that it had all apparently gone fine and
all I had to do now was to lie there with a lump of cotton
wadding covering one ear hallucinating *Californication*
Series 2 in sentence-sized bites. Every morning afterwards
I'd get up and get dressed. Then I'd sit on the chair in front
of my laptop. Not touching the keyboard, just sitting there
staring wonkily at the screensaver and the clock, hoping
that if I waited long enough the words and the colours might
mean something.

It was strange, but since the operation something had
happened to my sense of balance. My surroundings seemed

suddenly to have become animate. Nothing stayed where it was put. The floor rolled up at me and the walls rebounded. Trolleys leapt from the side of passageways and plates flew off tables into my face. I'd reach out to pick something up and find I'd misaimed – my hand was too close, or it was too far away, or it was a few inches to the right or left. I kept knocking into things, or find myself conducting a one-woman food fight. My legs and arms were splottered with bruises I didn't know how I'd got. Several times a day I'd find myself at the top of the stairs unable to figure out how they worked, or where I stood, or how exactly I was supposed to operate these extraordinarily complex pieces of spatial geometry. What are these things? Where do I put my foot? Where *is* my foot?

I tried everything I could think of to calm the Roaring Forties inside but none of it made much difference. I was drunk without drink, seasick far beyond sight of sea. In the evenings, when Simon had gone back to the hotel, I'd talk to the nursing staff, mugging up on my knowledge of French otology, or sit on the sofas in reception with the other stape-dectomy patients, bobbing at each other like overwintering flamingoes. '*Oh! Pardon, c'est à votre gauche. Moi? Oui, c'est la droite.*' Simon returned home and my sister Lucy came from London to see me. We sat in the town in pavement cafés wrapped in winter scarves, eating macaroons and gazing uncomprehendingly at the Christmas decorations. I kept trying to be myself but I couldn't grasp these new laws of misrule. Everything seemed so foggy. I was back in the gale, back on the boat, back in the dark with the lostness rising.

Lucy returned to England, and on the tenth day I wobbled down the corridor for the Grand Reveal. In the light of his office, Dr Vincent removed the surgical tape covering my ear and pulled away the wadding. I couldn't hear anything in that ear but that, he assured me, was normal – the eardrum needed time to heal.

Over on the other side of the building, the audiologist conducted another set of tests. She came round to my side of the booth with a tuning fork, banged it against the side of the chair and held it against the centre of my forehead. When this had been done after I first arrived, I'd felt the fork's vibrations equally on both sides. Now, I only felt them on the right. Before I'd been able to hear residual sounds in the left ear. Now, I could hear only a hollowness. Something, it seemed, had got lost in translation.

'Come,' said Dr Vincent, gesturing to the chair in front of his desk. 'Sit.'

When he looked at me, I saw that some of the wood-panelled distance between surgeon and patient had faded. There was a lot more of him in the room than there had been before.

'It is very unfortunate,' he said, 'but this is not the result that we wanted. The results of the audiogram show a drop in your hearing in the left ear.'

He seemed upset. Something had gone wrong that shouldn't have gone wrong and the wrongness had offended his sense of himself as a professional. Nought point five per cent or no, it was clear that Dr Vincent didn't like mistakes. I saw his offendedness and liked him the more for it.

'Has it gone?' I asked. 'Is it gone completely?'

'It is not the usual result,' he said. 'It is not right.'

'Can anything be done?'

'Right now, we will start you on a course of steroids.'

'Oh,' I said. 'I have to stay here?'

'Three days,' he said. 'You will need to stay for three more days.'

'Please,' I said, my heart dropping, 'I want to go home.'

'Let me try.' He was pleading too. 'I ask you – let us try another way. I am so sorry for this.'

I went back to my room. A nurse stuck a canula in the back of my hand. Another nurse arrived with a ligature and a tray of needles. When the nurses had left I turned over on my left, smothering the dying ear with the pillow.

The following day things had not changed. What I could hear through the left ear sounded like someone trying to whisper down the barrel of a flute – strange plasticky clankings with the meaning rinsed out. I was supposed to be in London at a meeting. A friend from Scotland was coming to stay this weekend. Another friend's mother who I had known since childhood had died but I couldn't be there at the funeral or the flat or the meeting because I was here with the canula and the barrel. But Simon, bless him, had returned.

After three days it was clear that steroids hadn't done the trick. 'I am so sorry,' said Dr Vincent. He looked stricken. 'I am more sorry because you are from London and it is not so easy for you to travel.' He waived his surgeons' fees and assured us that, should we choose to come back, he would do whatever he could to put this right. As he said goodbye he gave us a vexed, equivocal smile.

By the time we left the clinic it was late Friday afternoon. If we wanted to get the last ferry back to Dover on Saturday night, we'd have to be quick. We tried to rent a car, but it was a public holiday and all the cars were gone. *Please*, we said, *any car, any car at all, honestly. A bicycle. A donkey. A donkey with a puncture, even. Anything.* At the Hertz desk they looked at us (Simon, pulling a deck of cards from his wallet: RA, Tate, British Library … 'Um – do you take Oyster?' Me, ripping through the French phrasebook for 'Are there any significant changes in barometric air pressure between here and Calais which might in some way impact on recent otological surgery?') and gave us something to sign. We signed it. I drove. We got to a hotel. The hotel was all orange, a Hefneresque relic of the seventies. I don't know whether I dreamed or stayed awake.

In the morning, I drove and kept on driving while Simon

tried different kinds of music on the car stereo. Everything had an industrial quality as if it was being played from the back of a spaceship, all the words loaded with gravity. Before, deafness had been fog, but this was sound all fed back wrong, an acoustic mess. I couldn't position myself in a way to get the sound in and I couldn't find which way to turn. If I couldn't hear music, if all music was corrupted, then what then? Sod it, I said to Simon, sulky and melodramatic. I'd rather just gaffer-tape the stupid thing shut and stick a big notice on it saying, 'Gone.' Don't be daft, he said. Just wait.

A service station, somewhere north, rain sleeking. I was wearing the one-sided ear defender and as I stood by the fuel pump I watched the bandage around my hand come unfurled, spooling out in a long white flag. It fluttered in the wind till the rain pulled it down. The man at the pay booth looked at my one-sided headgear and as he gave me back my change, he leaned as far back in his chair as he could. Blimey, I thought, we must look like fugitives from justice.

Back in London. As time passed – Christmas, New Year, the middle of January – I noticed a change. Things in that left ear seemed to be getting louder. It was strange, but the previous week I'd been unable to hear planes on the Heathrow flight path and now there they were, obvious. How could I have missed them? And individual voices – Simon, my sisters, my mother, friends, colleagues – had begun to sound like themselves again. Something was going on, some kind of internal shifting. Was the ear healing itself?

One day in early February, the phone rang, and when I picked it up and put it to my left ear, I could hear my cousin's voice better than I could with my right even without the hearing aid. By spring, there was no question any more. The ear had sorted itself out. Not only had the distortion gone but the hearing in that ear was immeasurably better than in the right ear. I was astonished, grateful, wobbly. For a while I was wary of washing my hair or taking a shower in case my ear canal filled with water and drowned my good fortune. When someone called my name I turned my head carefully. Perhaps something vital might get rattled out of place. But every day the hearing was there, solid and real. Hear to stay. Each morning I could hear the world outside – radio, drills, sirens, train sounds. I'd lie there for a few minutes savouring the sheer pleasure of being woken by a mobile.

It took a certain amount of nerve to make the decision to go back to France for the second operation a year later. But this time, it was different. This time, Dr Vincent's surgery went without a hitch. I knew it was different as soon as the anaesthetic had worn off – no nausea, no rolling floors. When he took the wadding from the ear this time, the sound was raw, but it was – undeniably, absolutely – true sound.

A month or so after getting back I went to see Jacqui Sheldrake again. She did the tests and when she sat back, she beamed at me.

'Fabulous,' she said. 'Outrageously good.'

'You told me,' I said, 'that in the very best-case scenario I'd probably get back about eighty per cent of normal hearing.'

Jacqui prodded the fresh audiograms on her desk. 'It's more than that. This is just fabulous.'

Those head injuries, it turned out, had been red herrings. It had never been the skiing accident or the car crash which had caused my deafness, it had been the otosclerosis all along. The injuries may have accelerated its progression, but they had nothing to do with it – they'd just acted as decoys. If audiology is all otological detective work, then the clues had led directly to the wrong suspect.

'So basically,' I said to Jacqui, 'did I go deaf for twelve years unnecessarily?'

'No,' she said, 'definitely not.'

As she pointed out, the otosclerosis was there when I was first tested but it had been hiding behind the other symptoms. The cochlea won't submit to a photograph or an MRI so there was nothing that Jacqui or Steve or anyone else could do but put together the various bits of conflicting evidence, balance the sizeable probability that it was nerve damage, and come to the conclusion that the head injury had caused the problem. It had needed time – time to progress to the point where the clues were undeniable and the suspect stark.

The truth was, if I'd known, I probably could have had the operation a few years earlier. But if I had, I would have done so at the lowest point in that silence. As it was, I'd come up the other side. I'd discovered that deafness had its compensations and that there were ways of interpreting the world which had nothing to do with hearing. I had made – and found – my peace.

A FEW YEARS LATER, I ran into Jack Kartush. I say ran, but really it was the result of a series of coincidences so extravagantly far fetched as to make me wonder briefly if there might be something in this cosmic design idea after all. In a country far away, there was a man who introduced me to a man who told me over dinner there was a man he thought I should talk to. At the next table were a group of people sharing things with rum. Jack, said Barry, leading me towards a figure in the middle of the group, meet Bella. Bella, meet Jack. Jack does stapedectomies. In fact, he practically invented stapedectomies, and he certainly invented a lot of refinements to prosthetic stapes. Once I'd done with all the ohmygodthat'ssoamazings and ofallthelobstershacksetc, we fixed a time the following day to meet.

Jack … well, the only word which really works is 'trim'. Everything about him is trim. He's an economical sort of size, he's physically trim (late fifties, careful hair, engineered beard) and he exudes a sense of absolute delineation and containment – the way he rearranged his cutlery before he ate, the way he asked the waitress about each dish, the way he structured his sentences and lined the trajectory of his thoughts. I did not find it difficult to imagine him in an operating theatre laying out pieces of steel on a tray, and if his table manners are anything to go by, I assume he did surgery the way Fabergé did eggs.

Which is fortunate, since it wasn't until I talked to him that I truly understood the white-knuckle ultimacy of the extreme sport that is ENT surgery. First off – as Steve had pointed out long ago – the ear and the brain are effectively joined, which means it's impossible to be a surgeon of the ear without also being a neurosurgeon. Secondly, the ear is very, very well protected – better, in fact, than almost any other part of the body. And finally, for both those reasons, very few people ever get to meet it. The temporal bone is one of the hardest and most inaccessible in the human body, and thus the vast majority of medical students will pass through the full five to seven years of training without examining more than a two-dimensional textbook representation of a human ear. During cadaver dissection they'll see the brain, they'll see the spleen, they'll become intimate with the workings of the human heart, but none of them will ever actually see how we hear. Only those who then go on to specialise in otology will anatomise the real thing. 'Most people,' Jack points out, 'haven't the slightest idea of the complexity of the ear. Of course, the brain is complex and the heart is complex, but the inner ear has all that complexity, all that variability, within eighteen millimetres.'

Those who do qualify then find themselves drawn ever inwards. With heart surgery, there are several ways in and several ways out. With ear surgery, there's only one entrance – down the ear canal and through the skin at the side of the eardrum. Once in, they've entered a deeper chamber of secrets. 'The other name for the inner ear is the labyrinth, so it looks a bit like a maze. The cochlea looks like a snail.

So you have the coil for hearing, and then the three little balance canals, and wrapped tortuously right through that is the facial nerve, and next to that are the stapes, the incus and the malleus. And then the eardrum. And if you drill too far in the wrong area, you have the carotid artery that comes up from your heart to your neck to your brain, and if you enter too far into the carotid artery then the patient will exsanguinate on the table and die ...' – for some reason this really makes me laugh – '... and then just behind the carotid artery,' Jack continues, imperturbable, 'is the top of the jugular vein. Same thing.'

Just in case any of this seems a bit too workaday and straightforward, there're the potential knock-on effects of doing anything at all so close to the cochlea. Whatever movement the surgeon makes sets up a vibration which moves the cilia like little fields of wheat. 'You do this ...' he claps his hands '... and depending on the frequency, it will preferentially stimulate all those hair cells. So a small gesture is a huge tsunami for the inner ear.'

And, since this is all taking place in the immediate vicinity of various important bits of facial musculature and brain, there's less than minus room for error in the other direction as well. 'If you're an orthopaedic surgeon and you're doing big stuff, you're drilling a hole in a femur or an arm, then there's a little bone dust, you squirt a little water on it and then you put the screw in. But if you're doing microsurgery of the ear and brain, you have to do it under a microscope and if you're off by two mil, it's the difference between paralysing someone's face, or deafening them, or making them dizzy. Because the inner ear is one of the most sensitive organs of the body. If you think of sound as vibrating molecules, your ear has to pick up this nanometre vibration in order to be able to transduce it to your brain. So if you inadvertently go two mil too far into the inner ear when you're doing stapes surgery, if you catch an edge with the burr of the drill and spin off, then in a fraction of a second it's gone from where you want it to be to bumping into the facial nerve and paralysing your face.'

I consider this. Finally, I say, 'You must get good at breathing carefully.'

'Actually,' says Jack, 'that's true – and there are some moments with laser microsurgery where it's so critical that you have to watch the patient's in and out breath – when she exhales is the best time to shoot the laser.'

Thus, given his knowledge of just how high the stakes are in this particular game, what really puzzles him is why anyone would submit to the operation. 'I would be interested in examining why a person like you would put themselves under the knife when it also places you at risk for total deafness and dizziness. Why would you do that?' Because, I say, I'm sitting here surrounded by kids and voices and muzak in the background, and I can hear every word you speak. I understood that the risks were big, but I also understood that the potential benefits were bigger. My incentive is obvious. What's less obvious is why any surgeon should choose the anatomical equivalent of the North Face of the Eiger.

The clue, he says, is in the music. When he was growing up in Michigan in the 1960s he didn't want to be a doctor, he wanted to be a rock star, same as everyone else. At high school he had a band, grew his hair long, started writing his own stuff. When it became clear that stardom was unlikely he dropped out of the band, got his grades up and progressed to med school. Even so, and however far up he got, he always felt that he was 'a spy in the house of love, a bohemian long-haired guy pretending to be an academic'.

During a paediatrics placement, 'A nurse came in and said, "There's a kid come in with a torn outer ear." So I opened up my textbook and there was a German illustration, black and white, of the anatomy of the outer ear, the middle ear, the inner ear and the brain. And before I went in to see the child, I said, "This is beautiful, from both the left- and the right-brain perspective."' Though captivated by surgery, he never let go of the music but chose to spend his professional life getting up at 5 a.m., doing a full day's work, and then putting in his hours on the guitar. From the beginning he always understood that medicine and music have an entirely symbiotic relationship, and that the practice of one could enhance his skills in the other.

As a parting shot, he shows me a video of a stapedectomy. In terms of complexity it looked to me like an astronaut sewing buttonholes on the dark side of the moon, and if I hadn't understood before, I definitely understood then. It wasn't until the conversation with Jack Kartush that I realised just how lucky I'd been. How far beyond lucky, in fact. Out of the 11 million with hearing loss I had been one of the 2 per cent who had lost that hearing young, one of the one in a hundred who developed the clinical symptoms of otosclerosis, one of the tiny percentage of those who fitted the criteria for stapedectomy, and then one of the almost-nothing-sized club of people who got a perfect result from the operation. I was – I am – lucky beyond all imagining. I am bionic. I have two tiny titanium question marks in my ears, and I am rich beyond all the dreams of avarice.

13

Listening

SOUND HAD COME BACK into me with the force of revelation and I had no idea what to do with myself. I could hear! *I could hear!!!!* I'd been hearing for 28 years and deaf for twelve, and since I'd gone back to being hearing again, everything was bigger than I had the capacity to express.

I wanted everything. I wanted to try everything, listen to everything. I wanted to go up to strangers in the street and ask them if they had any idea of the miracles taking place inside their heads. I wanted to tell them that this hearing thing – this basic feature, fitted totally as standard in every working model – turned out, upon examination, to be a piece of kit which made the works of Shakespeare seem slack by comparison. I wanted to scroll dotingly through photos on mobiles, pull up proud scrapbooks of cochleas and temporal lobes, exchange reminiscences about auditory cortexes. I wanted to declare myself sound. I hoped these people knew how many miracles they had inside their heads, and just how much of the time they squandered those miracles on automated lift announcements and three-for-two offers on fabric conditioner.

I sat in cafés, blissed by the opportunity to eavesdrop on people bitching about their colleagues. I struck up conversations with strangers on trains or found excuses to offer directions to tourists. I rang up friends in Orkney or Greenock just because I wanted to hear the way they said 'modern' or 'cosmetic' and savoured the tastes of each professional dialect – the wipe-clean tones of nursing staff or get-in-quick diction of cold-callers. Several times I lost the thread of discussions because I was too busy listening to the sensation of listening rather than the sense. I talked to people on the tube. I took my new hearing to films, parties and bicycle races, I experimented with power tools and hung out round chainsaws. I stood below telegraph lines to hear the scribble of swallows or climbed hills to find the lilt of a curlew. I greeted the three-note preamble to a train announcement like an old friend and tripped out on the sheer poetry in 'Cashier number THREE, please!' I watched TV not because I was interested in what was on, but because I loved the indulgence of sitting there just moving the volume button up and down. I wasn't groping for a single word any longer or making approximate swipes at possible topics. I could hear a whole sentence! Every letter of every word! I could make out all of what people were saying from beginning to end! I was astounded by the thrill of exactitude. I could hear accent, dialect, nuance, mood. I could understand, and once I understood, I could connect. I had come home.

And that, to be honest, is how it felt. For 28 years I had been a native in the land of the hearing and for twelve I had been a traveller through the world of the deaf. Some time in 1998 I had taken my place in the 10-million-strong queue of dispossessed hearers all shuffling down the lines towards a place we didn't want to be. I'd left protesting and bewildered, and I'd arrived at my destination transfigured. I'd been one person and now I was another. Somewhere along that journey I'd discarded many of the things I'd been carrying and picked up new tools better suited for the job. I'd left a lot of myself behind, and I'd brought some along for the ride. I learned the language, I picked up a few local habits and secrets, I met some people I would be glad to call friends for life. By the end, I loved the deafened world and understood it as a place of magnificence and revelation. But the truth is, I had never stopped walking. I was always just a tourist.

At the same time as I was savouring sound I was also readjusting to a world in which I could, if I chose, be completely indifferent to it. In many ways I found it bizarre how easy I found it to return to the world I'd lived in before. There was nothing I had to do, no recalibration to be made. Sound was there, sound was gone, sound came back. It was as if I'd walked back into the space I'd walked out of twelve years ago and found it unchanged. Two stupendously sophisticated, complex operations had produced something that was ... simple.

For a couple of months afterwards, I still dabbed around on the bedside table in search of the hearing aids every morning when I woke up. After the operations I'd put the aids away in my make-up bag, where they sank beneath a layer of eyeliners and vanished from view. After a year or so – or some period of time long enough to convince myself this new hearing thing was for real – I took them out, cleaned them up and sent them to Jacqui so she could plunder the mechanisms for someone else. When the second ear had healed fully, I also discovered I was no longer tired all the time. Because my brain was no longer working at full capacity to filter and process sound, whole holds of internal storage space seemed suddenly to have become free. I didn't need nine hours of sleep a night, and I no longer slept like I'd been hit. And if I was woken by car alarms or drills, well, that seemed like a fair exchange to me.

I also felt a certain amount of survivor's guilt. I had been astonishingly lucky, and I knew it. For twelve years, I had believed that this was only going to get worse, and then at the last moment I had been offered an alternative. At the moment, there are very few hearing conditions which are operable, though the possibilities continue to expand every year. The real breakthrough will come when we can work out how to regenerate hair cells just as birds do. After all, if there are 11 million people with hearing loss in the UK alone, then that's 11 million incentives to improve the situation. Until then there are no real cures for sensorineural hearing loss, only remedies.

So for all I had returned to a world I'd inhabited before, this time it was different. Though it might initially have seemed reasonable to behave as I had for the first 28 years, it wasn't the same. I knew more. I understood more. I understood what hearing could do and what it couldn't and the spaces it could fill between one person and another. I understood that I had been given a second chance, and that it was my job to live every last drop of that chance. So I got happy – just straightforwardly, normally happy. I moved out of London, kept writing, used what I'd learned about listening and about life. The things I'd discovered while deaf came with me. This time around I truly knew the value both of what I'd got, and what I'd got back.

And one day it might happen again. Some stapedectomies last for ever, some don't, and statistically I'm exactly as likely to suffer age-related hearing loss as anyone else. But I'm lucky. If it happens again, I know the old country now. I know its landscape, something of its politics and a lot of its people, and if I need to, I can go native.

ONE OF THE THINGS I did pick up on my travels was a genuine fascination for sound. My friend's old charge – 'You're not deaf, you just don't listen!' – had fallen away, and what was left in its place was something else. Not a sense of judgement or waste, but a deeper understanding of what hearing really meant.

At school, I had come across the writings of someone called Tony Parker. He was a Quaker and a pacifist who had become a conscientious objector during the war and a prison visitor after it. His work in jails suggested to him that locking people up was never going to be a solution to the problem of crime, but understanding why people got locked up just might be. His singular mission in life was to listen to people, and by listening to give them a voice. In doing so he invented a whole new literary form – neither oral history nor conventional to-and-fro interview but a hybrid of the two. Though the majority of his work was on criminals and criminology, over forty-odd years of writing he published books on everything from Belfast during the Troubles to a tiny town in the middle of Kansas. His only unifying theme was marginalisation. He didn't talk to people who were famous or who were used to being interviewed. He talked to squaddies or people who lived on South London council estates or men serving life sentences for murder, and his interviews are as much a record of the simple power of listening as a testament to the abundance of human life.

Look Parker up and he appears as an oral historian, but he wasn't, not really. Oral historians are after something – they're chasing parts of the past, or they're looking at a person through a specific depth of focus. But Parker wasn't interested in people for the sake of history. He was interested in people for the sake of them. He didn't care about the preservation of a legacy. All he cared about were the things within people which remain eternal. And, though the dialect of crime shifts so fast the voices in his interviews often sound quaint, behind the gor-blimey-gov patter, he's talking to the same people you probably talked to yesterday.

Parker always started with a brief introductory paragraph describing the person and their immediate surroundings and then left the rest to them – up to ten pages of transcript during which each interviewee told their own story completely in their own words. Parker's working methodology was scrupulous: he would usually visit the person three or four times and though he would guide the conversation he'd also let it wander where it wanted to go. Once he felt he had got enough material he would begin transcribing, taking out the ums and ahs and smoothing out the narrative snags but never altering a single word. If he needed to make things clearer, he made sure he patched in a phrase from a different part of the interview. Idioms, region and accent all come across clearly, and if the details may be dated, then the human nature behind them very definitely isn't. Everything that appeared in print had been said by the interviewee, and if they didn't want to be identified then he didn't identify them. People liked and trusted him instantly, whoever they were and whatever they'd done, because they understood that he didn't judge them. He was known as the Great Listener.

Parker had heard those people with an open heart and no agenda. He was never afraid to ask difficult or troubling questions, and there aren't many people doing what he did. He was a one-off, but he may also have been a bit of a prophet. He understood that to properly listen to someone – drop everything, sit down, forget everything except the person in front of you and what they're saying – is an act of communion. People would talk to him not just because he asked the right questions but because nobody else asked any questions at all. It didn't matter whether he was dealing with a company chairman or the mother of a dozen children. The majority of the people he spoke to were starving: starved of the opportunity to reveal themselves and be listened to, starving for contact. He came along with his mild shoes and his forgettable style and there they were, laid out for him like pages. He could, I suppose, have exploited them. Instead, he took what they said and made it sacred.

It's a platitude now that we're all supposed to be obsessed with communication, but though the pace and quantity of that communication have increased, it doesn't mean that the quality has improved. Half of us spend our days on hold while the other half sits in call centres with a script in their hand and a clock by their eyeline. The companies and organisations we're trying to get through to aren't listening because they're not interested in us, they're interested in our money. And our data, because that leads to more money. Public services are suspicious of something they think we've done, or they want to know how we're going to help our kids achieve their Key Stage 4s. Politicians want our love in vote form, counsellors want us for the experience, gurus want our souls, and nobody's going to tell the truth at work. Why would we? There's far too much at stake.

Even back at home, all our histories just get in the way. Once upon a time, perhaps you thought your wife's views were interesting. Now she talks, and you don't even register she's speaking. Besides, everyone has moments when they choose to hear selectively. Children – who usually have very acute hearing – are brilliant at sorting out useful noises (a Fifa game) from superfluous ones (siblings), and most people would probably recognise the phenomenon of marital shell-shock, in which someone can be medically deaf to a request to cook dinner but spring-loaded to the sound of a bottle-opener three doors down.

But there's a danger somewhere in failing to listen. In the same way that a human can be starved of love or of touch, so they can be deprived of connection. If someone spends years feeling like everything they'd sent has never been received, then how would they start to behave? If they felt that the louder they shouted the greater the returning silence, how would that alter the way they saw the world? Would they keep trying to communicate, or would they just give up?

When I was working in Edinburgh during my twenties, I had a friend who wasn't conventionally good-looking but who was still impressively successful with women. His success attracted awe and envy from his male friends and puzzlement from those female friends who had known him too long for whatever-it-was to work.

'Go on,' I said to him one evening, watching a couple of senior editors scrapping for his favours. 'Tell me.'

'Easy,' he said. 'All you have to do is look a woman in the eyes and give her your complete and undivided attention.'

'No,' I said. 'Really.'

'Honestly,' he said. 'All I do is just sit there and listen. I don't think about anything else, I just concentrate on her. Fifteen minutes max. Bad day, twenty.'

And he was right. Over the next couple of months I watched him at it: direct eye contact, total focus, fifteen minutes, sold. I was impressed – we all were, in a horrified sort of way. It was a fine demonstration of the power of attention, but there was also something disconcerting about it. Were people really so unused to being heard that fifteen minutes was all it took?

So who is doing what Tony Parker did now? Who is listening with an open heart and no agenda? Shrinks, counsellors, psychologists; the usual suspects, paid or unpaid. Plus probably a few people you wouldn't expect. On a black cab ride a few years ago, the driver told me that he regularly took confessions. He wasn't a priest – he had moved to London from Poland 27 years earlier, married a Japanese woman and had three daughters, all of whom now have PhDs. During the day he developed property and at night he drove a cab just for the pleasure of it. What did he enjoy so much? The people, of course, he said. If you like people, cab-driving is a fantastic job. Plus there were other benefits.

'You wouldn't believe what people do in taxis.'

'Oh,' I said. 'Like what?'

'People getting down on their knees like in church, shuffling up to the partition and talking through that little cash slot. Asking for forgiveness.'

'No *way.*'

'I swear! Because they know I'm a stranger and that they'll probably never see me again. I'm sitting here looking at the road in front and all they can see of me is a pair of eyes and they just want someone they can say things to.'

'How often?'

'Oh,' he said, 'you'd be surprised. Regular. Once every few weeks at least.'

'So what is it they say?'

'Men want to brag,' he said. 'They're drunk or they're high and they want to tell you all about the deal they've just done. Or the affair they're having and the women they want.'

'And the women?'

'They want a shoulder to cry on. About men, mostly. Regrets. Kids. How to save their dead marriages.'

Most times I've taken a cab since – any city, any cab – I've asked other drivers for their experiences. Yes, they said, almost unanimously. Yes, you get people confessing, and yes, they want absolution. Half of taxi-driving is driving, and the other half is psychiatry. A city isn't made of brick and pavements, a city is made of people.

And it's not just cab drivers who take our tales. Many GPs would say that their waiting rooms are crowded not with patients in need of pills or hospital appointments, but people in search of someone to whom they can tell all the hurts in their heads. Hairdressers, nail technicians, spa therapists – ostensibly they're there to help with dodgy cuticles or lower back pain, but actually they're there to act as human shields against loneliness.

The trouble is that listening is like caring – one of those nebulous metaphysical things which doesn't submit willingly to performance indicators or stakeholder outcomes and which is therefore unlikely to be provided free (or otherwise) by the NHS. Which means that where it's offered at all, it's outsourced to the usual bundle of charities, religious organisations and multi-competent middle-aged ladies. Both the UK and the USA have any number of different helplines and organisations devoted to specific causes – domestic violence, immigration issues, hearing voices, being Christian, being the victim of a forced marriage, being depressed, being Chinese and depressed ... Several churches around the country now offer 'Crisis Listening' services for those in need of some kind of secular absolution, and there's a whole further series of organisations specifically set up to deal with children's issues. Given that the first helpline was set up in 1953 and that before then the only other options were friends, family or the local priest, something huge seems to have happened to the place and status of hearing in the intervening years.

If – as Oliver Headley and the army had proved – sound is a thousand times more powerful than we give it credit for, then so too must be the power of being heard. Most of us are used to the idea of using song or dance to alter our mood, but less so to the idea that just being listened to is itself a harmony and a balm. Everyone has things they probably don't want to hear: their kids' fighting, stuff about debt or divorce, the news from Syria. The trouble is that the logical endpoint of all that blocking is pseudohypacusis, the deafness with no apparent biological cause which Dr Sally Austen had been talking about, in which both body and mind conspire to ensure that whatever it is the individual can't face hearing is not physically admitted. But there's a big difference between offering someone a better connection and knowingly taking on another man's poison. If you completely listen then you completely open yourself. Which is when all the interesting things start to happen.

14

Music

IF YOU ASK PEOPLE what they miss when they lose their hearing they'll usually mention specific voices or songs. And then after that, they'll almost invariably talk about birdsong. It doesn't seem to matter how much they actually know about birds or how connected they feel to nature. They could be lifelong twitchers or they might have difficulty telling a pigeon from a pterodactyl, but what they mind is the absence of that loveliest, most inconsequential of sounds.

Between birds, every note presumably carries as much meaning as words do among humans, but because we can't understand that meaning, all we can hear is the whole. It's entirely possible that all we're really listening to is the avian equivalent of 'did you take the bins out?', but it doesn't matter – all that really matters is its existence. For them, it's a biological imperative, but to us it just sounds like pure, naked joy.

Similarly, it's not necessarily the full symphonic splendour of a dawn chorus in the country that people need most, but urban birdsong. During daylight hours the magpies and parakeets are there as reminders that there's more to a city than just humans, but after dark on a January night blackbirds enchant with their lonely solos from satellite dishes and power lines, from the tops of shop signs and the ends of cranes. That's what we need, and that's what we miss – not just voices we understand, but all those high-up reminders of sound's unsayable loveliness.

ABOUT SIX WEEKS after getting back from the second operation I had gone with an old friend to a concert. Miranda
had somehow got returns to see the Berlin Philharmonic
under Sir Simon Rattle. The Berlin Phil are famous for their
creamy sound and their many-splendoured reinterpreteta-
tions of the classics, and although they may be too rich (in
all senses) for many people, the chance to see them is not the
sort of thing you turn down. I can't remember what was on
the programme – Haydn? Schubert? – but it wasn't music I
knew. The right eardrum was still half healed and I shouldn't
really have gone, but I'd taken my old hearing aids to wear as
earplugs just in case.

We shuffled programmes and leaned over the seat-rests
for a quick catch-up. Lights down. The well-trained hush
of the audience. Then a clatter of applause as Rattle walked
towards the conductor's podium. He stood, faced the orchestra, raised his baton …

And then something happened.

Until that moment, my experience of the world had been
scanned through a digital filter. For the past twelve years,
everything had been met at the door and given a quick electronic frisking before being sent on its way to my brain.
Every sound, every note, every word had first had to get past
the little plastic gatekeepers I kept in my ears. But on that
February evening when the first notes rushed towards me,
instead of meeting the usual wall of electronic resistance,
they met nothing. Nothing except openness. And, meeting
nothing, the music broke straight through. At this time and
in this place, it filled me from tip to brim. It swept in through
my senses and danced in my brain, smashed open my heart
and blew the bloody doors off my lovely, precious, astounding hearing. I sat there, sound-blasted, while a few bars of
Schubert changed everything.

The music poured in, a great shining river, pounding like
a waterfall over every atom in every corner of my being. And

as it swept through I could feel, genuinely feel, neurons and synapses sunk for years into darkness snap back into life. It was as if someone had walked in to my engine room, found the internal fuse box and with one great downward slam of their hand had thrown both switches – *Bang!! Bang!!* – and now every light in the place blazed out again. I walked into that concert hall monochrome, and I left it in colour.

It was the first true music I'd heard in more than a decade, and I promise you, I absolutely promise, that if you should ever have cause to question the power of sound or its capacity to reset the very cells of you, then try going deaf and then getting your hearing back after twelve years. Science had given me back my hearing, and now music had returned me to life. It blew everything, transcended everything. It's astounding. It was a thousand volts of birdsong, a blackbird translated.

All I had known during all the time that I was deaf was that there was something missing, something I couldn't put my finger on. Whatever that thing was, it wasn't just connected to the straightforward mechanical facts, it was something invisible, something that slipped in unnoticed alongside a tone of voice or a harmony, something which took sound from being a mono or stereo process to a multidimensional comprehending. I'm struggling to explain it now, so there was no way I could have taxonomised it at the time. I couldn't have said, look, I know I can hear the notes all right, but the song seems to be missing. But at that moment, in that place, with an old concerto I'd never listened to before or since, I understood. Music is as powerful as it gets. It is love, made liquid.

I think maybe the whole thing was almost worth it just for that single experience. Almost.

Further Reading

Chapter 3: AID
Mälzel and Beethoven's ear trumpets at the Beethoven Haus, Bonn:
http://www.beethoven-haus-bonn.de/sixcms/list.php?page=
ausstellungsstuecke_museum_en&skip=10
History of hearing aids: http://theinstitute.ieee.org/technology-
focus/technology-history/the-history-of-hearing-aids

Chapter 4: LOSS
Statistics on hearing loss: https://www.actiononhearingloss.org.uk/
your-hearing/about-deafness-and-hearing-loss/statistics.aspx
NHS Action Plan on Hearing Loss, Prof. Sue Hill, Kevin Holton,
Cathy Regan, Department of Health, March 2015: https://www.
england.nhs.uk/wp-content/uploads/2015/03/act-plan-hearing-
loss-upd.pdf
Hearing loss and mental health: https://www.actiononhearingloss.
org.uk/~/media/Documents/Policy%20research%20and%20
influencing/Research/Mental%20health/Mental_health_report.
ashx
Deafness in Mind, Sally Austen and Susan Crocker, Whurr
Publishers, London, 2005
Pseudohypacusis: False and Exaggerated Hearing Loss, James E.
Peck, Plural Publishing, San Diego, 2011
Psychogenic deafness: http://onlinelibrary.wiley.com/
doi/10.1111/j.1600-0447.1954.tb05696.x/abstract
Occupational Hearing Loss, Robert Thayer Sataloff, Joseph Sataloff,
3rd edn, Taylor & Francis, Boca Raton, FL, 2006
'The treatment of hysterical deafness at Hoff General Hospital',
Andrew I. Rosenberger and James H. Moore, *American
Journal of Psychiatry*, http://ajp.psychiatryonline.org/author/
ROSENBERGER%2C

Chapter 5: CONDUCTION
Beethoven's symptoms: http://hearinghealthmatters.
    org/hearinginternational/2011/
    hearing-beethoven-part-ii-the-medical-conclusion/
Letters of Beethoven, vol. I, Emily Anderson, Macmillan, London,
    1961, LL
Beethoven: Biography of a Genius, George R. Marek, William
    Kimber, London, 1970, LL
Beethoven Depicted by His Contemporaries, Ludwig Nohl, trans.
    Emily Hill, W. Reeves, London, 1880
The Life of Ludwig van Beethoven, Vol. 3, Alexander Wheelock
    Thayer, CUP, 1921, Google Books
Beethoven: Anguish and Triumph, Jan Swafford, Faber & Faber,
    London, 2014, LL

Chapter 6: ROCK
Music and volume: WHO factsheet on hearing loss: http://www.
    who.int/mediacentre/factsheets/fs300/en/
German study on musicians' risk of hearing loss: https://www.bips-
    institut.de/no_cache/en/publications/press/single-view/artikel/
    laerminduzierter-hoerschaden-vierfach-hoeheres-risiko-bei-
    professionellen-musikern.html
French horn players: http://www.telegraph.co.uk/news/health/
    news/10334527/French-horn-players-are-most-at-risk-of-
    hearing-loss-in-an-orchestra.html
Classical v. rock risk of loss: https://www.cnet.com/news/
    shocker-the-most-deaf-musicians-arent-rockers/
Musicians with hearing loss: http://www.musicianshearingservices.
    co.uk/category/testimonials/
George Martin *Telegraph* obituary, 9 March 2016: http://www.
    telegraph.co.uk/music/artists/george-martin-the-greatest-
    music-producer-who-ever-lived/
George Martin hearing loss: http://www.jazzwax.com/2012/09/
    interview-sir-george-martin-pt-4.html
Giles Martin: http://studio.sonos.com/article/conversation-giles-
    martin

Chapter 7: ACOUSTICS

Build the Ships: The Official Story of the Shipyards in Wartime, V. S. Pritchett, MoI, 1946, HMSO, LL

Enquiry Into the Effects of Loud Sounds upon the Hearing of Boilermakers and Others Who Work Amid Noisy Surroundings, Thomas Barr, Royal Philosophical Society of Glasgow, 1886: https://archive.org/details/b21457384

The Diseases of Occupations, Donald Hunter, English Universities Press, London 1962

An Inevitable Consequence: The Story of Industrial Deafness, Dick Bowdler, https://www.dickbowdler.co.uk/content/publications/The-Inevitable-Consequence.pdf

Noise, Rupert Taylor, Penguin, London, 1975

Restaurant acoustics: Andrew Corkill, Director, Spectrum Acoustics: email 9 August 2016

'How much noise can we stand?', Rupert Taylor, *Daily Telegraph Magazine*, no. 346, 11 June 1971

Chapter 9: DISTORTION

Noise and Noise-Induced Hearing Loss in the Military, ed. Larry E. Humes, Lois M. Joellenbeck and Jane S. Durch, National Academies Press, Washington, DC, 2006: https://www.nap.edu/read/11443/chapter/5#78

Research on Hearing Loss, US Dept of Veterans Affairs: http://www.research.va.gov/topics/hearing.cfm

Use of hearing protection on military operations: https://www.ncbi.nlm.nih.gov/pubmed/22319982

Impact of Noise on Hearing in the Military, Jenica Su-ern Yong and De-Yun Wang, Military Medical Research, 2015: http://mmrjournal.biomedcentral.com/articles/10.1186/s40779-015-0034-5 (UK)

NIHL payments to US vets: *War Is Loud: Hearing Loss Most Common Veteran Injury*: http://backhome.news21.com/article/hearing/

Conductive Hearing Loss – Synopsis of Causation, S. S. Musheer Hussain and Kevin Gibbin, MoD, September 2008: https://www.gov.uk/government/publications/synopsis-of-causation-conductive-hearing-loss

Sensorineural Hearing Loss – Synopsis of Causation, Dr Adrian
Roberts and Kevin Gibbin, MoD, September 2008: https://www.
gov.uk/government/uploads/system/uploads/attachment_data/
file/384550/sensorineural_hearing_loss.pdf

'Heritage of Army audiology and the road ahead: the Army
Hearing Program'. *Am. J. Public Health*, 98(12), December
2008, pp. 2167–72: http://www.ncbi.nlm.nih.gov/pmc/articles/
PMC2636536/ inc. Table 2: 'Effect of tank crewmen's ability to
understand spoken orders on their performance in combat
situations in the US Army'

Ministry of Defence FOI response 2014: https://www.gov.uk/
government/uploads/system/uploads/attachment_data/
file/346766/PUBLIC_1408522542_armed_forces_with_
hearing_loss.pdf

Ministry of Defence FOI response 2016: https://www.gov.uk/
government/uploads/system/uploads/attachment_data/
file/508823/Defence_Statistics_FOI_2016_02860____The_
number_of_UK_Armed_Forces_personel_graded_from_H1_
to_H4_including_noise_hearing_loose_claims_from_2011_
to_2015.pdf

Military hearing loss – online forums: https://www.arrse.co.uk/
community/threads/hearing-loss-can-i-claim.196099/
http://www.leatherneck.com/forums/showthread.php?54482-
va-benefits-for-hearing-loss/
http://www.pprune.org/military-aviation/488748-hearing-loss-
armed-forces-compensation-scheme.html

Irish army compensation claims: *Whistleblower, Soldier, Spy: A
Journey into the Dark Heart of the Global War on Terror*, Tom
Clonan, Liberties Press, Dublin, 2013

Looking for Trouble, General Sir Peter de la Billière, HarperCollins,
London, 1994, LL

'Welcome to "The Disco"', Clive Stafford-Smith, *Guardian*, 19 June
2008: https://www.theguardian.com/world/2008/jun/19/usa.
guantanamo

'Music as torture: war is loud', David Peisner, *Spin*, 30 November
2006: http://www.spin.com/2006/11/music-torture-war-loud/

'When music is violence', Alex Ross, *New Yorker*, 4 July 2016: http://
www.newyorker.com/magazine/2016/07/04/when-music-is-
violence

'What America learned about torture from Israel and Britain',
Joshua Keeting, *Slate*, 15 December 2014: http://www.slate.com/
blogs/the_world_/2014/12/15/what_america_learned_about_
torture_from_israel_and_britain.html

Sound in the Age of Mechanical Reproduction, ed. David Suisman
and Susan Strasser, University of Pennyslvania Press,
Philadelphia, 2010: https://books.google.co.uk/books?id=
UCbeD5RvI_oC&pg=PA2&lpg=PA2&dq=shafiq+rasul+noise&
source=bl&ots=Nq56ksnjqP&sig=DejXak92CuaI_VuJple7JGXo
NG4&hl=en&sa=X&redir_esc=y#v=onepage&q=shafiq%20
rasul%20noise&f=false

Sound torture during the Iraqi War: Justin Caba, 20 January 2015:
http://www.medicaldaily.com/torture-methods-sound-how-
pure-noise-can-be-used-break-you-psychologically-318638

US Senate Intelligence Committee Report 2014: https://www.
intelligence.senate.gov/sites/default/files/documents/CRPT-
113srpt288.pdf

Chapter 10: SIGN

Far From the Tree, Andrew Solomon, Vintage Books, London, 2012

Chapter 11: VISION

Inside Job: Lucian Freud in the Studio, Photographs by David
Dawson, Hazlitt Holland-Hibbert, London, 2004

Chapter 12: SURFACING

Soldier, Soldier, Tony Parker, Heinemann, London, 1985
The People of Providence, Tony Parker, Eland, London, 1983

Acknowledgements

THIS BOOK TOOK A SHORT TIME to write but a long time to make. While I was researching it I spoke to many people about sound and hearing, and though I wasn't able to include all of them, their involvement was vital. Katrina Burton, Catriona Hetherington, Oliver Searle, Professor David McAlpine, Rupert Taylor, Charlie Swinbourne and Caroline Parker all gave their time and insights without reservation. I'm also immensely grateful to all those I have quoted: Jacqui Sheldrake, Dr Robert Vincent, Professor Victor Humphrey, Steve Rakkar-Thomas, Adam Sieff, Andy Hearn, Sir Peter de la Billière, Oliver Headley*, Giles Martin and Jack Kartush, Dr Sally Austen and everyone at the Birmingham & Solihull Deaf Mental Health Service (names of patients are pseudonyms for privacy's sake, but you know who you are).

Heartfelt thanks to Cecily Gayford at Profile for her forbearance through several lively bouts of editorial catherding, to Kirty Topiwala for offering the full splendour of Wellcome's research resources, to Andrew Franklin, Anna-Marie Fitzgerald and Valentina Zanca at Profile for making the whole process such good fun, and to Simon Garfield for a revision I followed to the letter. I'm also grateful to Professor Alan Palmer for fact-checking over his Christmas break, and to Victoria Hobbs for holding fast.

But the biggest thanks of all goes to those who were there all the way through: Lucy, Flora, Rupert and Shera Bathurst,

---

* Not his real name.

Alex, Ruth and Danny Renton, Kamal Ahmed, Ruaridh Nicoll, Paddy and Petra Cramsie, Miranda Holt, Clare Hulton, Tony Daniels, Angus and Stephanie Wolfe-Murray, and Urièle and Marc. 'Thanks' is far too small a word. 'Love' covers it better.